Eco-Friendly Coatings and Adhesive Technology

Eco-Friendly Coatings and Adhesive Technology

Guest Editors

Guangpu Zhang
Zhengmao Ding
Yanan Zhang

Basel • Beijing • Wuhan • Barcelona • Belgrade • Novi Sad • Cluj • Manchester

Guest Editors

Guangpu Zhang
School of Chemistry and
Chemical Engineering
Nanjing University of Science
and Technology
Nanjing
China

Zhengmao Ding
Pen-Tung Sah Institute of
Micro-Nano Science
and Technology
Xiamen University
Xiamen
China

Yanan Zhang
College of Materials Science
and Engineering
Nanjing Tech University
Nanjing
China

Editorial Office
MDPI AG
Grosspeteranlage 5
4052 Basel, Switzerland

This is a reprint of the Special Issue, published open access by the journal *Polymers* (ISSN 2073-4360), freely accessible at: www.mdpi.com/journal/polymers/special_issues/183ZX651E1.

For citation purposes, cite each article independently as indicated on the article page online and using the guide below:

Lastname, A.A.; Lastname, B.B. Article Title. *Journal Name* **Year**, *Volume Number*, Page Range.

ISBN 978-3-7258-2758-9 (Hbk)
ISBN 978-3-7258-2757-2 (PDF)
https://doi.org/10.3390/books978-3-7258-2757-2

Contents

Review

Research Progress of Self-Healing Polymer for Ultraviolet-Curing Three-Dimensional Printing

Wenhao Liu [1], Zhe Sun [1], Hao Ren [1], Xiaomu Wen [2], Wei Wang [3], Tianfu Zhang [3], Lei Xiao [1] and Guangpu Zhang [1,*]

[1] National Special Superfine Powder Engineering Research Center of China, School of Chemistry and Chemical Engineering, Nanjing University of Science and Technology, Nanjing 210094, China; 13586987884@163.com (W.L.); sunzhe97529@163.com (Z.S.); rhmoon1112@163.com (H.R.); 15005161138@163.com (L.X.)

[2] Science and Technology on Transient Impact Laboratory, No. 208 Research Institute of China Ordnance Industries, Beijing 102202, China; wenxm2908@163.com

[3] Science and Technology on Aerospace Chemical Power Laboratory, Hubei Institute of Aerospace Chemotechnology, 58 Qinghe Road, Xiangyang 441003, China; wang7469061@163.com (W.W.); charlestfzhang@sina.com (T.Z.)

* Correspondence: gpzhang@njust.edu.cn

Abstract: Ultraviolet (UV)-curing technology as a photopolymerization technology has received widespread attention due to its advantages of high efficiency, wide adaptability, and environmental friendliness. Ultraviolet-based 3D printing technology has been widely used in the printing of thermosetting materials, but the permanent covalent cross-linked networks of thermosetting materials which are used in this method make it hard to recover the damage caused by the printing process through reprocessing, which reduces the service life of the material. Therefore, introducing dynamic bonds into UV-curable polymer materials might be a brilliant choice which can enable the material to conduct self-healing, and thus meet the needs of practical applications. The present review first introduces photosensitive resins utilizing dynamic bonds, followed by a summary of various types of dynamic bonds approaches. We also analyze the advantages/disadvantages of diverse UV-curable self-healing polymers with different polymeric structures, and outline future development trends in this field.

Keywords: self-healing; UV-curing; 3D printing; polymer

Citation: Liu, W.; Sun, Z.; Ren, H.; Wen, X.; Wang, W.; Zhang, T.; Xiao, L.; Zhang, G. Research Progress of Self-Healing Polymer for Ultraviolet-Curing Three-Dimensional Printing. *Polymers* **2023**, *15*, 4646. https://doi.org/10.3390/polym15244646

Academic Editor: Annalisa Chiappone

Received: 24 September 2023
Revised: 27 November 2023
Accepted: 4 December 2023
Published: 8 December 2023

1. Introduction

3D printing technology has been widely used since the 1980s in many fields, such as automobile manufacturing [1], the aerospace industry [2] and biomedicine [3]. The high RTM (resolution/time for manufacturing ratio) of 3D printing technologies which use ultraviolet light as a curing source makes it popular for practical applications [4]. Molecules containing light absorption units, namely chromophores (such as C=C, C=O), are in an excited state after being irradiated by a certain wavelength of ultraviolet light. The excited-state molecules undergo a crosslinking reaction and change from the original liquid state to the solid state. The principle of UV-curing 3D printing technology is to use ultraviolet light to selectively cure the photosensitive resin under the control of digital signals. The liquid resin is attached to the previous curing layer and then cured. The process is repeated until a complete printed product is formed [5]. UV-curing molding technologies include stereolithography (SLA) [6], digital light processing (DLP) [7], liquid crystal display (LCD) [8], and so on. A working principle diagram of several photocurable 3D printing technologies is shown in Figure 1, and their advantages and disadvantages are compared in Table 1. The stable cross-linking networks make prepared products difficult to recycle and reprocess when they generate microcracks and defects. At the same time, volumetric shrinkage is unavoidably generated as the polymerization reaction proceeds, and the van der Waals forces between the monomers are converted into covalent bonds, resulting in a shorter distance between the monomers [9,10]. Based on these factors stated above, the accuracy and mechanical properties of the prepared products are reduced. Therefore, some researchers have introduced dynamic bonds into polymer chains to realize the self-healing function of the materials. The dynamic "breaking-recovering" reversible process of bonds helps to release intramolecular shrinkage stress produced during the polymerization process, thereby reducing the volume shrinkage [11,12].

Self-healing materials can repair their defects and restore their original performances under certain external stimuli (light or heat). Up until now, a variety of self-healing materials have been successfully prepared and used, such as concrete [13,14], metallic [15,16], and polymer materials [17,18]. Self-healing polymers are widely used in biomedicine [19,20], protective coatings [21,22], and other fields [23]. Based on the different self-healing mechanisms, such polymers are divided into extrinsic [24] and intrinsic types [25]. Extrinsic self-healing polymers are prepared by filling the raw material resin with micro-/nano-structures such as micro-capsules [26] and micro-fibers [27] containing healing agents. For instance, Shinde [28] and Sanders [29] et al. prepared a UV-curable resin containing self-healing micro-capsules, achieving good 3D printability and a satisfactory self-healing effect. However, the number of repairs was limited because the healing agent was exhausted when the crack increased [30,31]. In contrast, the healing process in intrinsically self-healing polymers is based on the fracture and recombination of dynamic bonds in the polymer chains. Dynamic exchange reactions are activated by external stimuli and can theoretically be exchanged infinite times. Therefore, intrinsic self-healing polymers have attracted a great deal of attention. Dynamic bonds are divided into dynamic non-covalent bonds (including ionic bonds [32], hydrogen bonds [33], coordination bonds [34], etc.) and dynamic covalent bonds (including Diels–Alder bonds [35,36], ester exchange bonds [37], disulfide bonds [38], imine bonds [39], etc.). Compared with dynamic non-covalent bonds, dynamic covalent bonds usually have higher bond energy [40] and endow materials with better mechanical properties. Thus, to realize self-healing, a high temperature or energy is necessary [41], which causes some restrictions on the application of dynamic covalent bonds.

In summary, it is highly desirable to develop photosensitive 3D printing resins with a self-healing ability. At present, UV-curable dynamic-bond polymers have been intensively investigated, and studies aim to introduce a self-healing function into UV-curing 3D printing products. Many dynamic bonds, including disulfide bonds, ester exchange bonds, Diels–Alder bonds, hydrogen bonds, ionic bonds, and metal coordination bonds, have been used to construct UV-curable 3D printing materials. This paper mainly introduces the common dynamic bonds in UV-curing 3D printing polymers and evaluates the performance advantages and applications of different materials.

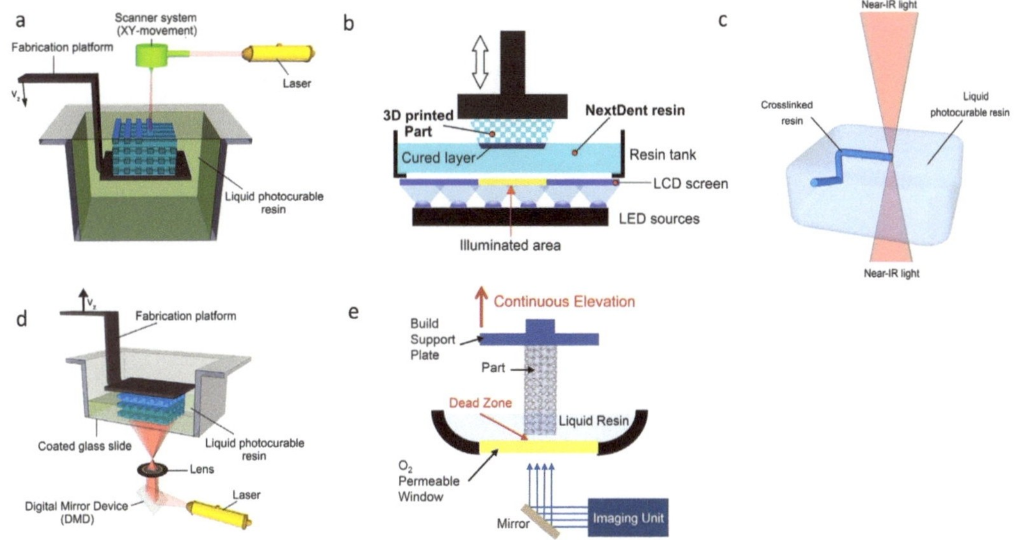

Figure 1. (**a**) Schematic diagram of stereolithography (SLA) [42]. (**b**) Schematic diagram of liquid crystal display (LCD). Reprinted with permission from reference [43], 2021, ELSEVIER. (**c**) Schematic diagram of two-photon 3D printing (TPP) [42]. (**d**) Schematic diagram of digital light processing (DLP). Reprinted with permission from reference [42], 2012, ELSEVIER. (**e**) Schematic diagram of continuous liquid interface production (CLIP). Reprinted with permission from reference [44], 2015, SCIENCE.

Table 1. Comparison of various popular UV-curing 3D printing technologies [45,46].

Name	Operation Principle	UV-Curing Mechanism	Advantage	Disadvantage	Application
SLA	Laser beam single-point printing	Free radical and hybrid curing	Mature technology, form large size device	Slow curing speed	Dentistry, mold, automobile
DLP	Projection printing	Free radical curing	Fast curing rate, high precision	Form small size device	Medical care, jewelry, education
CLIP	Projective continuous printing	Free radical and thermocuring	Extremely fast curing speed	Expensive resin and equipment	Sports, cars
MJP [1]	Multi-nozzle printing	Free radical and hybrid curing	High precision, colorfulness	Expensive equipment	Consumer goods, medical care jewelry
TPP	Dual Laser Beam Printing	Free radical curing	Extremely high precision	Expensive equipment, complex process	Microelectronics, art, scientific research
LCD	Liquid crystal imaging printing	Free radical curing	Fast curing speed, low cost	Short service life	Jewelry, mold manufacturing

[1] MJP [47] technique is also called PolyJet.

2. Dynamic Covalent Self-Healing Polymers for UV-Curing 3D Printing

Due to the different types and properties of dynamic bonds, polymers with different properties can be constructed. Under certain external stimuli, these polymers achieve restoration by rearranging their dynamic covalent bonds. In general, dynamic covalent bonds exhibit high bond energy and stability, which endow the polymer with a higher mechanical strength and modulus.

2.1. Dynamic Disulfide-Bond

The relatively low bond energy of the disulfide bond (about 250 kJ·mol⁻¹) [48] makes it easily activated at room temperature [49] or by a certain intensity of UV-light irradiation [38]. Under the action of free radicals or anionic intermediates, disulfide bonds are activated and then recombined to achieve self-healing. The disulfide bond exchange reaction is shown in Figure 2.

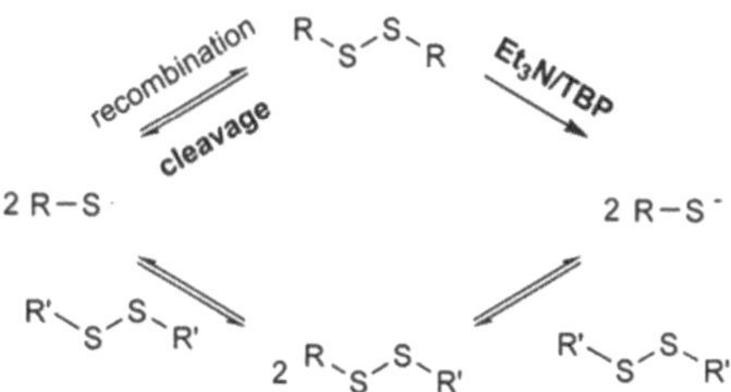

Figure 2. Disulfide exchange through the formation of sulfur-based radicals and sulfur-based anions. Reprinted with permission from reference [50], 2016, ELSEVIER.

Li et al. [51] reported a UV-curable polyurethane acrylate resin (PUSA) based on disulfide-bond self-healing for DLP 3D printing. The authors first reacted bis(2-hydroxyethyl) disulfide and polyethylene glycol with isophorone diisocyanate to prepare an isocyanate-terminated prepolymer (PUSA). Then, the isocyanate group was terminated with 2-hydroxyethyl acrylate. The prepared PUSA showed a mechanical self-healing performance under UV irradiation. The tensile strength and elongation at break were 3.39 ± 0.09 MPa and $400.38 \pm 14.26\%$, respectively. After healing at 80 °C for 12 h, the spline could withstand large tensile deformation, and the healing efficiency reached 60% of the initial value. Even after healing for the third time, the tensile strength of the samples reached 2.05 ± 0.34 MPa.

The formula for self-healing efficiency is as follows:

$$\eta = \frac{\sigma_{healed}}{\sigma_{original}} \tag{1}$$

In the formula: $\eta(\eta_{stress}/\eta_{strain})$—tensile strength (stress/strain) repair efficiency; σ_{healed}—tensile strength after repair; $\sigma_{original}$—original tensile strength. A series of control experiments confirmed the decisive effect of the disulfide bond on the self-healing. The prepared UV-sensitive resin exhibits moderate viscosity, good fluidity, and a fast UV-curing rate. The double-bond conversion rate after 60 s of UV irradiation is about 80%. Moreover, it shows good compatibility with commercial DLP 3D printers and can be used to print various 3D products. The printed products have high definition, smooth surfaces, and preserved mechanical and self-healing functions. Owing to its excellent performance, such a polymer has great application potential in the fields of UV-curable coatings, adhesives, and inks.

Thiolene photopolymerization is a much more efficient way to crosslink elastomers than the copolymerization of double bonds with acrylate monomers [52]. So Yu et al. [53] reported a UV-printable self-healing elastomer based on sulfhydryl and disulfide bonds. Under UV-light irradiation, the polymerization of sulfhydryl and alkene [54,55] activated rapidly. The metathesis reaction of disulfide bonds in the system endows the elastomers with a self-healing function. It is worth noting that the number of sulfhydryl groups affects the UV-curing rate, while the number of disulfide bonds affects the healing performance. This research on the interaction between the UV-curing speed and healing efficiency paved the way for a subsequent similar system [56], in which a balance between two parameters was achieved. In addition, the ratio of the cross-linking groups to the self-healing groups can be regulated via the oxidation reaction of the thiol group, which, in principle, facilitates the preparation of polymers with the desired mechanical performance and self-healing function. Thiolated (mercaptopropyl) methylsiloxane-dimethylsiloxane (MMDS) was oxidized by Iodobenzene diacetate (IBDA) [57] to form oligomers containing both thiol and disulfide functional groups (Figure 3a). Then, the oligomers were reacted with vinyl-terminated polydimethylsiloxanes (V-PDMS) with a double bond to form a solid elastomer. V-PDMS has a relatively low viscosity (below 200 cSt), which is suitable for the stereolithography process. The printed elastomers have a high resolution (up to 13.5 μm) and excellent self-healing performance. After healing at 60 °C for 2 h, the apparent cracks reconnected and the tensile strength was restored (Figure 3b). High healing performance is still achieved even after more than 10 cycles of self-healing tests (Figure 3c); however, the Young's modulus of the experimental elastomer is relatively low (17.4 kPa). The research on the theoretical model of self-healing guides the understanding of self-healing polymers and the use of dynamic bonds [58–60].

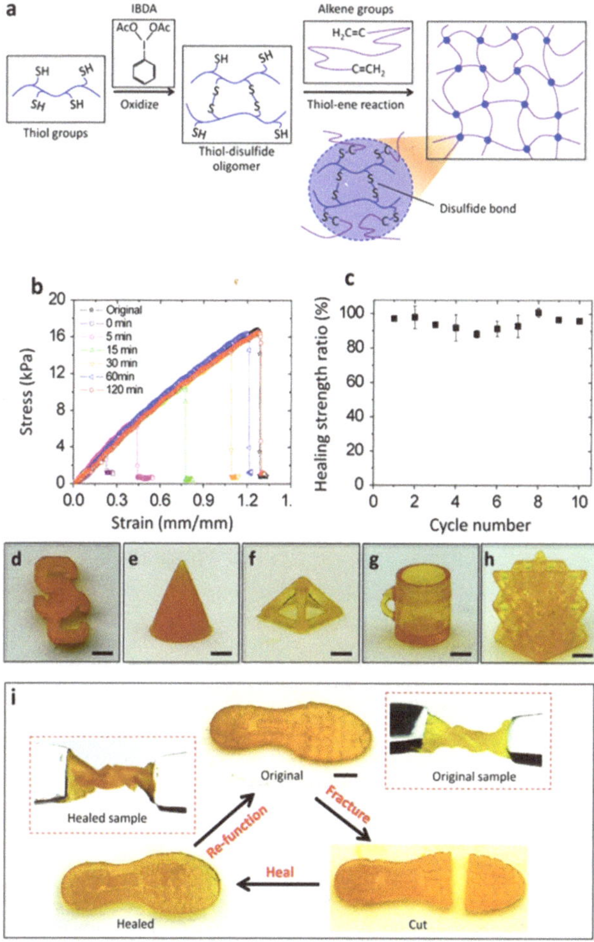

Figure 3. (**a**) Molecular design of the self-healing elastomer. (**b**) Nominal stress–strain curves of the original and self-healed experimental elastomers for various healing times. (**c**) Healing strength ratios of the experimental elastomers for 10-cycle healing tests (each 2 h at 60 °C) (**d–h**). The manufactured samples. (**i**) Self-healing of a shoe pad sample (scale bars: 4 mm). Reprinted with permission from reference [53], 2019, Springer Nature.

Zhang et al. [38] prepared a chemically cross-linked ionic gel (ionogel) system with the synergistic effect of dynamic disulfide bonds, hydrogen bonds, and ion-dipole interaction. The synergistic effect of various forces is beneficial to enhancing the material's performance. Polyurethane acrylate with disulfide bonds was synthesized and then compounded with the diluent monomer 4-Acryloylmorpholine, the ionic liquid, and a photoinitiator to obtain a UV-sensitive resin. The UV-cured elastomer shows good transparency and can be modified by adding different color additives. By adjusting the mass ratio of the raw materials, the tensile strength of the ionogels can reach 7.42 MPa, and the elongation at break is 977%. The mechanical performances of the ionogels remained after continuous cyclic tensile/compression tests. The micro-pressure sensors were prepared by DLP3D printing technology. The sensors have excellent elasticity and durability, and good response sensitivity and self-healing functions. The healing efficiency after UV irradiation for 10 min is greater than 99%. The results confirm that the ionogel has great application prospects in wearable ionotronics and intelligent soft robotics.

Gomez et al. [61] presented a UV-curable, self-healing elastomer system that can achieve ultra-high elongation (1000%). The elastomer realized the modular manufacturing of soft robots by DLP printing technology. The acrylic elastomers have chain and reticular structures that can be adjusted by adding small amounts of dithiol and dynamic thioether crosslinkers. Through a thermally induced reverse Michael reaction, the free thiol and acrylate portions rapidly combine to form new thioether bonds for effective polymer diffusion and self-healing. The polymer has a complete self-healing ability in multiple damage/repair cycles via dynamic disulfide-bond exchange. These elastomers are compatible with self-healing stretchable electronic devices. It was used for 3D printing and integrated to form highly complex, large-scale functional soft robots. The presence of dynamic bonds can eliminate the need for adhesives or complex connectors during the connection of the sub-components. This is of practical significance for simplifying the component process.

2.2. Transesterified

Ester bonds are widely found in a variety of commercial thermosetting polymers, such as epoxy resins, unsaturated polyester resins [62], alkyd resins [63], etc. As shown in Figure 4, the hydroxyl and ester groups contained in the polymers can undergo a rapid dynamic transesterification reaction [64] at high temperatures, which does not change the topology of the polymer, and provides it with good self-healing performance.

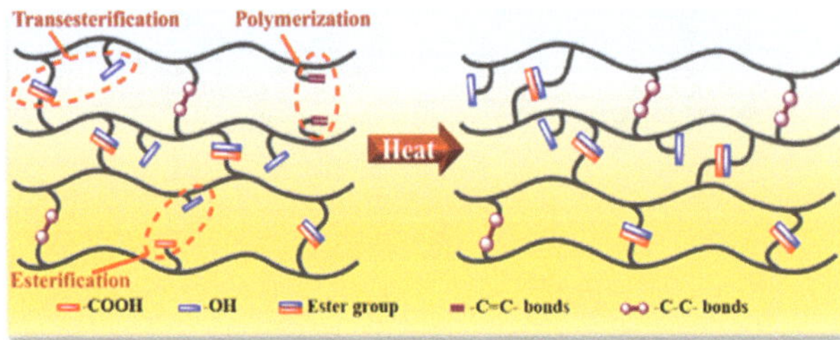

Figure 4. Illustration of the repair mechanism based on the transesterification. Reprinted with permission from reference [64], 2021, ELSEVIER.

Grauzeliene et al. [37] used 2-Hydroxy-2-Phenoxypropyl acrylate and bio-based acrylate epoxidized soybean oil to form a new type of vitrimer that can be DLP printed. The rich hydroxyl and ester groups in the system provide the necessary conditions for the transesterification reaction. The printed product has a two-way shape memory, 47% self-healing performance, 31% re-processability, and 100% alcohol degradation recovery performance. This product was used as artificial muscles and actuators; however, its high healing temperature (200 °C) and re-processing temperature (180 °C) limit its applications. The same authors designed and synthesized a vitrimer based on glycerol and vanillin for environmental protection strategies [65]. Owing to the high rigidity and thermal stability of the main chain of vanillin, the vitrimer exhibits excellent performance. Acrylated epoxidized soybean oil was first used in UV-curable resins for the synthesis of vitrimer. The use of environmentally friendly soybean oil as a raw material is in line with today's green development concepts.

2.3. Imine Bond

The dynamic exchange reaction of imine bonds is also used in the field of self-healing [66,67]. The thermally initiated imine exchange reaction is mild and occurs rapidly without considerable side reactions. The imine bonds can recombine at room temperature [68].

Liguori et al. [69] prepared a vanillin-based UV-curable resin with self-healing ability that can be used as a DLP printing material and can be mechanically and chemically recovered by heat treatment. The Schiff base reaction between the aldehyde functional group of vanillin and the amino group of ethylenediamine generates an imine functional group. The hydroxyl functional group of vanillin reacts with acrylic anhydride to obtain a UV-reactive double bond (Figure 5a). As a result, the cured thermosetting material exhibits a high fracture strength (17.3 ± 3.9 MPa), extensibility, self-healing, and thermal re-processability due to the presence of a cross-linking structure and imine bonds. In addition, it can be chemically recovered via transamination in ethylenediamine (Figure 5b), and the obtained oligomeric product with amine end groups can be used to produce novel thermosetting films. Tensile tests showed that the elastic moduli of mechanically and chemically recycled thermosetting plastics are similar, whereas the elastic moduli of self-healing samples are slightly higher and the elongation and fracture stress are slightly lower. This resin was demonstrated to be suitable for producing 3D objects via DLP printing (Figure 5c,d), although the printing accuracy is not as good as that obtained using non-biological resins [70].

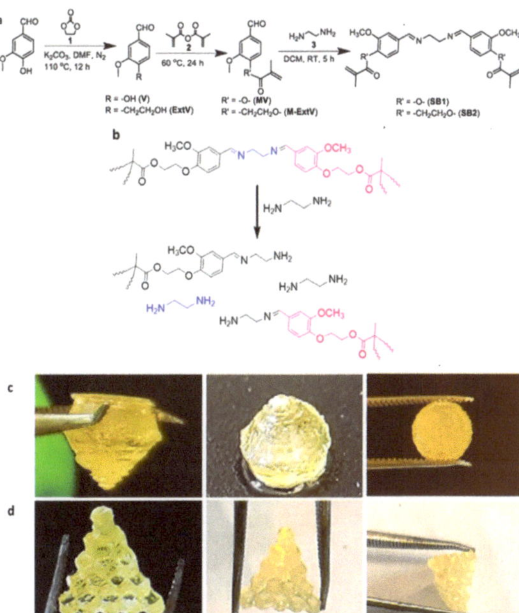

Figure 5. (**a**) Scheme of reactions for the synthesis of resins. (**b**) Suggested transamination pathway leading to the solubilization of resin in ethylenediamine. (**c**) 3D-printed diamond; (**d**) 3D-printed stairs. Reprinted with permission from reference [69], 2021, ELSEVIER.

To explore the effect of different Jeffamines on the properties of cured products, Cortes-Guzman et al. [39] adjusted the type and chain length of the polyamines, and five thermosetting polymers with different mechanical performances were obtained. The Young's moduli of these range between 2.05 and 332 MPa. They were applied to various applications based on their different performances. Since triamine provided a higher cross-linking density network than the diamine, the former was found to show a higher ultimate tensile strength (UTS) and Young's modulus; however, the degree of freedom of the polymer chains decreased and the elasticity was inferior to that of the diamine system. Because of the simple synthesis process and commercial availability of Jeffamines, this series of resins provides a promising alternative to commonly used printing formulations. Furthermore, such bio-based polymers are expected to replace those made from nonrenewable resources [71].

2.4. Diels–Alder Bond

The reversible Diels–Alder addition reaction occurs between electron-deficient diolefins and electron-rich conjugated dienes at low temperatures and the reverse reaction is activated at high temperatures [72]. It is widely used in the field of self-healing polymers. A representative example is the reaction between furan and maleimide [73,74]. Structures containing a Diels–Alder bond network exhibit higher intermolecular force and better mechanical performances than non-covalent bonds [75,76]. Until now, the reversible Diels–Alder reaction has been scarcely applied in the field of 3D printing, because it requires a high temperature to initiate healing.

Durand-Silva et al. [35] studied the effect of thermally reversible Diels–Alder cross-linking agents on the shape stability and self-healing performances of UV-printable resins. Aiming to optimize the ratio of raw materials to provide a self-healing function (>99%) without affecting the shape stability of the printed products, the authors prepared a UV-sensitive self-healing resin using acrylic furan−maleimide (fm) cross-linking agent and 2-Hydroxyethyl acrylate (2-HEA) as a reactive diluent (Figure 6a).

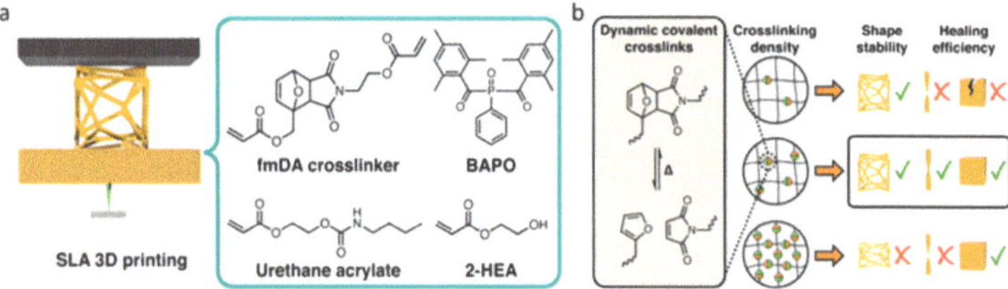

Figure 6. (**a**) Formulation of self-healing resins. (**b**) By varying the concentration of dynamic cross-links in resins, it is possible to balance the self-healing efficiency and the shape stability of printed objects. Reprinted with permission from reference [35], 2021, American Chemical Society.

The results showed that the extent of the effect on the 3D printing shape increased with the concentration of the dynamic covalent cross-linking agent (Figure 6b). The healing efficiency did not show a linear relationship with the concentration of the covalent cross-linking agent, and no self-healing phenomenon was observed even at a high concentration, in contrast to general beliefs. Similar to other self-healing printing polymers, reliable self-healing polymers based on furan−maleimide Diels–Alder (fmDA) adducts must strike a balance between shape stability, thermal stability, and dynamics.

3. Dynamic Non-Covalent Self-Healing Polymers for UV-Curing 3D Printing

The activation of dynamic bonds usually requires conditions such as a high temperature or energy input [77], which involves additional energy use. The bond energy of dynamic non-covalent bonds is often lower than that of covalent bonds. The polymers formed through dynamic non-covalent bonds are less susceptible to the external environment and, therefore, require lower-intensity stimulation to heal the damage [78].

3.1. Hydrogen Bond

Hydrogen bonding is a common physical interaction that is much weaker than covalent bonds. When hydrogen atoms are adjacent to highly electron-rich atoms such as N, O, or F, reversible cross-linking networks are established via electrostatic interactions in polymer chains. Increasing the temperature can weaken the hydrogen bonding force, which is conducive to the movement of polymer chain segments. When the temperature decreases, the hydrogen bond will be regenerated again [79]. The hydrogen bond is suitable for constructing intrinsically self-healing materials and, thus, merits considerable research attention.

Wu et al. [80,81] selected a series of UV-curable methacrylic acid and acrylic monomers as inks to apply in UV 3D printing. A vinylated palm oil monomer (POFA-EA) was selected as the raw material for the printing of thermoplastic polymers. The ink was obtained by blending N-vinyl-2-pyrrolidone (NVP) and acrylic acid (AA) (Figure 7a), and various biobased thermoplastics were successfully printed using LCD technology. The ink exhibits a high double-bond conversion rate of 73–85% within 8 s, which meets the polymerization requirements for LCD printing. The amide structure in the vinylated palm oil monomer forms hydrogen bonds with polar monomers, including NVP and AA, enable the transparent printed products to exhibit high stretchability (Figure 7c) and self-healing abilities (Figure 7b), as well as easy processing and recycling abilities. The authors also explored the printability of a series of UV-curable monomers such as N-hydroxyethyl acrylamide (HEAA), AA, NVP, etc. The formation of strong hydrogen bonds between the –NH and C=O groups of polymer chains allows the polymers to meet the LCD requirements. Even mixing with other monomers does not affect the printability of N-hydroxyethyl acrylamide. This study guided the selection of the appropriate printing monomers. The viscous elastomer prepared in this paper can adhere to human skin, enabling repeated stripping–adhesion after 10 cycles without leaving any residue or causing skin irritation. These excellent features were exploited to prepare biosensors and wearable devices, demonstrating the great potential of this elastomer as a biomedical material.

Using natural tannic acid, choline chloride, and hydroxyethyl methacrylate as raw materials, Zhu et al. [33] developed novel ternary polymerizable deep eutectic solvents (PDESs) using the one-pot method. Density functional theory analysis indicated that hydrogen bonds and van der Waals interactions were the main driving forces for the formation of PDESs. The prepared PDESs showed high bio-based content (40.13–52.96%) and excellent mechanical performances (tensile stress, 5.45–13.71 MPa; toughness, 13.40–23.29 MJ·m^{-3}). The degree of healing reached 85.2–88.5% after heating at 80 °C for 24 h, and the performances after healing were still superior to most reported PDES materials [82–84]. Because tannic acid exhibits natural UV-light absorption [85] and anti-bacterial properties, these new PDESs can be directly used for LCD printing without additional inhibitors [86]. The penetration depth of the PDES sample was low (0.192 mm), and the different structures that were printed all showed high resolutions. The penetration depth and critical exposure energy [87] of UV-sensitive resin were investigated, which paved the way for the further improvement of printing accuracy. Overall, the developed tannic acid-based PDESs can serve as ideal materials for high-resolution 3D printing, and are equipped with both green and multifunctional properties.

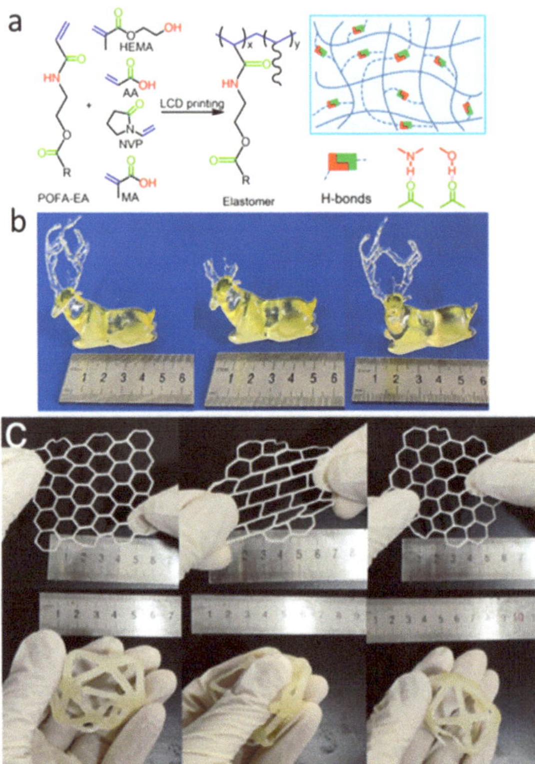

Figure 7. (**a**) Reaction mechanism of PO-based elastomers formed using POFA–EA and comonomers (HEMA, AA, NVP, or MA). (**b**) Photographs demonstrating self-healing in 3D deer model. (**c**) Reversible stretching and deformation of 3D objects. Reprinted with permission from reference [80], 2021, American Chemical Society.

Li et al. [88] synthesized a deep eutectic solvent-based UV-curable resin with both hydrogen bonds and ionic interaction by mixing choline chloride (ChCl), acrylamide (AAm), and 4-Acryloylmorpholine (AcMo) (Figure 8a). The resin has an ultra-low viscosity (<0.1 Pa·s) and ultra-high curing rate, which is compatible with commercial LCD printers. The printed glassy components have high precision (50 microns) and optical transparency. The synergistic hydrogen bonds between hydroxyl, amino, and carbonyl groups in the polymer chains provide self-healing properties (75.9%). This polymer shows an extremely high stiffness and humidity-related conductivity, rendering it suitable for manufacturing humidity-responsive devices. In addition, two-dimensional structural assemblies with different sizes and their solubility in recycling and re-modeling were demonstrated in Figure 8b–d. The presence of choline chloride contributes to the solubility of the printed products, enabling sacrificial mold manufacturing. It is convenient for the manufacturing of precise, multifunctional structures on demand.

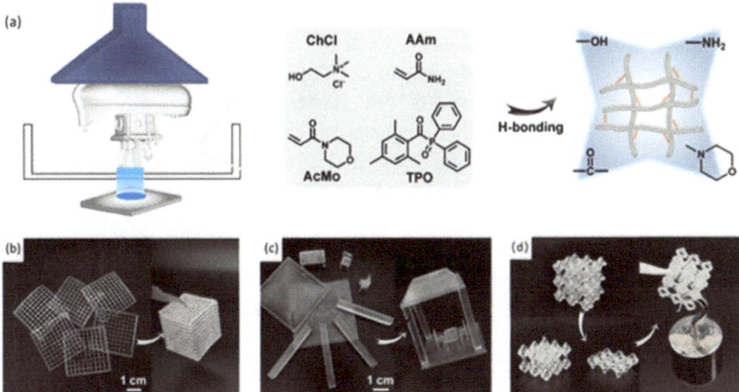

Figure 8. (**a**) LCD printer for the fabrication of 3D objects, the composition of the resin, and the hydrogen-bonding network in the resin. (**b**) 2D gridding to a 3D-box structure; (**c**) small parts to a complex pavilion-like building. (**d**) Photographs showing the self-healing capability of a 3D-printed lattice. Reprinted with permission from reference [88], 2023, American Chemical Society.

Zhu et al. [89] reported a dynamic polymer with highly customizable mechanical performance. Owing to the ionic and hydrogen bonds inside the polymer, the DLP-printed samples have good self-healing and recycling abilities. In addition, the printed products were easily tuned from soft elastomers to rigid plastics by adjusting the ratio of urethane monoacrylate/acrylic acid monomers. The on-demand preparation of complex structures, various assembly categories, and healable functional devices was realized by using this polymer. To a certain extent, this approach solves the environmental problems caused by traditional DLP thermosetting products. Soft and rigid models that were printed showed a good recovery effect in shape and strength after healing at 90 °C for 12 h. The printed products were recovered via thermocompression to obtain uniform samples. The tensile strength recovery rate, fracture strain recovery rate, and Young's modulus recovery rate were determined to be 92%, 85%, and 109%, respectively. After recycling them multiple times, a 70% restoration of the mechanical performances was still observed.

Wu et al. [90] synthesized a solid conductive ionic gel (SCIg) with a rapid curing rate and self-healing function. A precursor solution composed of acrylic acid and choline chloride was mixed with a gelatin cross-linking agent containing many amino acid residues in the molecular chain. The acrylic acid and choline chloride formed a polymer network with inherent conductivity and the remaining amino acids generated a second rigid network, which allowed for the preparation of a new type of SCIg with excellent comprehensive performance. The rich hydrogen bonding sites in the system enhance the non-covalent interaction and afford excellent mechanical performances. Moreover, the addition of glycerol enhances the physical cross-linking networks and improves the healing ability (>95%). Based on the conductivity of the polymer, a stretchable resistive sensor, a compressible capacitive sensor, and an electronic skin were constructed by DLP, demonstrating the application of SCIg in flexible wearable devices. This SCIg achieves a balance between ionic conductivity and mechanical performance. Problems such as solvent leakage and the evaporation of gel elastomers containing liquid can be avoided by using SCIg. In addition, its rapid curing rate and high printing accuracy make it highly promising for application in the fields of electronic skin, physiological signal detection, and human–machine interfaces.

Invernizzi [91] and Suriano [92] et al. developed an SMP that was obtained via the crosslinking reaction of diol-terminated polycaprolactone (PCL) with 2-isocyanate ethyl methacrylate and methacrylate monomers. The PCL segment provides a shape memory function, and the quadruple hydrogen bond established by ureidopyrimidinone provided self-healing performance. Experiments proved that the DLP-printed 'L'-type specimen shows similar tensile strength to the cast specimen The ureidopyrimidinone moiety does not harm the mechanical performances. The healed samples are still suitable for applications such as soft actuators.

Durand-Silva et al. [93] synthesized methacrylate monomers with aliphatic or aromatic urea motifs, which can exert a good dilution effect on oligomers. The presence of different types of hydrogen bonds improved the toughness and self-healing ability of the products. The cross-linking density increased with the increase in the urea group content. By controlling the content of hydrogen bonds, the mechanical strength was increased by 119% and the toughness was increased by 205%. They studied the effect of pendant aliphatic and aromatic ureas on the mechanical performances of polymer networks. As described in previous reports [94], they found that the presence of pendant hydrogen-bonded urea enhanced the mechanical tensile strength and toughness without reducing the elongation.

In the hydrogel system that is rich in hydrogen bonds, the mechanical strength and self-healing ability of the polymer can be further enhanced by introducing ion interactions, and the resulting conductivity can broaden the application scope of the polymer.

Huang et al. [32] developed a thermoplastic polymer composite with self-healing (>60%), efficient recyclability, and a customizable mechanical performance for DLP. The polymer contains soft and hard regions. The introduction of zinc methacrylate provides the composite with a second dynamic bonding (ionic bonding). The soft mono-functional urethane chains and ionic bonds suppresses the brittleness by acting as energy absorption units between the hard 4-Acryloylmorpholine segments. The complementarity of hard and soft chain segments considerably improves the mechanical strength of the component and the self-healing performance, thus facilitating the combination of the components with different mechanical properties. This is important for the re-assembly or recycling of robotic components after failure. Stemming from the superior combination of self-healing polymers, it outperforms most reported self-healing polymers, with a tensile strength and elastic modulus of up to 49 and 810 MPa, respectively. This study introduces self-healing and printing functions into a polymer without affecting its recyclability and strength, equilibrium self-healing ability, or mechanical performance.

Wu et al. [95] proposed an inter-penetrating network hydrogel based on poly(acrylic acid–N-vinyl-2-pyrrolidone) and carboxymethyl cellulose. They were physically cross-linked through Zn^{2+} coordination and hydrogen bonding. The printed hydrogel exhibits the same dynamic dual-physical interactions as 'sacrificial bonds' [96], resulting in a high tensile toughness (3.38 $MJ\cdot m^{-3}$) and good self-healing ability (η_{stress}, 81%; η_{strain}, 91%). The hydrogel was subjected to 40%-strain tensile cycles 100 times, and resistance responsiveness was still observed. The hydrogel can also be customized for flexible sensor printing. Owing to the designability and high resolution of its 3D printing (the profile is still smooth and complete at a minimum scale of 100 μm), the hydrogel has broad application prospects and avoids the limitations of the hydrogel-extrusion-based printing technology. In terms of resolution and printing speed, this hydrogel is expected to prepare wearable, flexible sensors.

3.2. Crystallization

Crystallization [97,98] is a process in which atoms or molecules are arranged according to certain rules to form crystals. Generally, polymers do not form regular single crystals owing to their long chain structure and chain entanglement behavior. However, polymers with ordered structures can form crystal embryos, from which crystals can eventually grow as the ordered structure cools from the amorphous state. Above the melting temperature (Tm) of the crystal, the crystalline polymer flows. When the polymer is damaged, a segment can flow to the damaged place. Upon the temperature decreasing, the crystal is re-formed. Therefore, crystalline polymers can also be self-healing.

Zhang et al. [99] incorporated a semi-crystalline linear PCL (relatively low melting temperature of 60 °C) with good miscibility on a system of SMP methacrylate (Figure 9a). After heating the prepared spline above 60 °C, the PCL crystal domain melts and the PCL linear chain is diffused through the boundary between two separated samples. When cooled to room temperature, the PCL linear chains formed crystalline domains. Some crystal domains cross the boundary between individual strips and combine them to achieve a repair effect (Figure 9b,c), which constitutes a solution to the irreparable thermosetting polymers. The prepared dual-network SMP exhibits a UV-curable self-healing ability (self-healing shape memory polymers, SH-SMP) and good compatibility with DLP 3D printing. This polymer is suitable for manufacturing complex printing structures with high resolution (up to 30 μm) (Figure 9d,e). The advantages of the system were demonstrated by printing some samples, including an SH-SMP holder and a cardiovascular stent. The potential application prospects of this system in the fields of soft robots, flexible electronics, and biomedical equipment were proven.

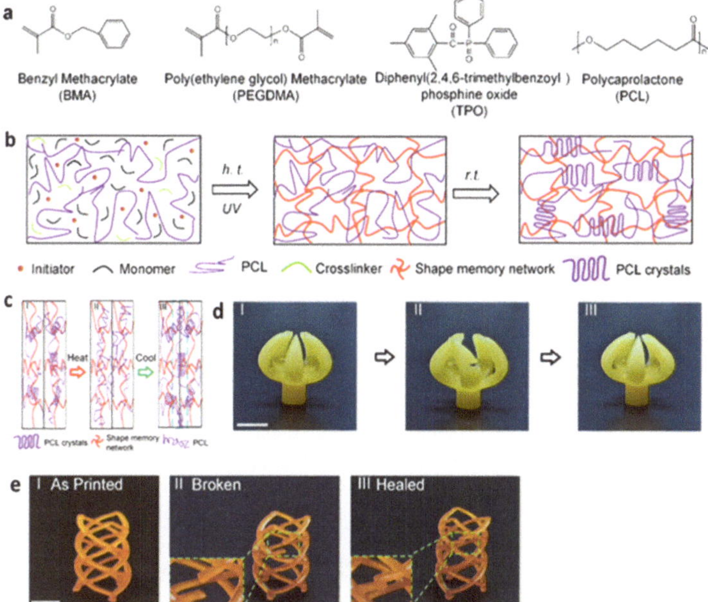

Figure 9. (**a**) Chemical structures of the components in SH-SMP solution. (**b**) Chemical structural evolution of the SH-SMP solution during UV-based 3D printing at high temperature (h.t.) and cooling down to room temperature (r.t.). (**c**) Illustrations of the self-healing mechanism. (**d**) 3D-printed gripper. (**I**) As-printed gripper. (**II**) Cut gripper. (**III**) Healed gripper. (**e**) 3D-printed stent. (**I**) As-printed. (**II**) Broken. (**III**) Healed (scale bars: 4 mm). Reprinted with permission from reference [99]. 2019, American Chemical Society.

Wen et al. [100] mixed PCL with a UV-curable monomer to print a semi-interpenetrating polymer network elastomer with shape memory, self-healing function, and high tensile strength (20 MPa). The appearance of the sample after cutting and repairing was the same as the original, and the shape recovery speed decreased only slightly.

Abdullah et al. [101] proposed a temperature-responsive hydrogel. The hydrogel consists of hydrophilic poly (acrylic acid) chains containing different molar fractions of the hydrophobic segment C16A. The physical cross-linking of co-polymer chains can be achieved via hydrophobic association between the crystalline domains of hydrophobic segments. The melting and crystallization temperatures of the hydrogels are 38–40 °C and 25–29 °C, respectively. It allows for inducing a shape-memory function at temperatures close to the human body by optimizing the C16A content. By adjusting the molar fraction of the hydrophobic segment, the printed hydrogel can be made to exhibit a Young's modulus of 23–215 MPa and toughness up to 7 MJ·m^{-3}, enabling a transition from brittleness to toughness. The developed 4D-printable hydrogel has great potential in various biomedical applications.

3.3. Host-Guest Interaction

Host–guest interactions [102] are mainly caused by hydrophobic interactions and complementary shape and size characteristics between the host and guest molecules. The dynamic interaction between the host and guest molecules can endow the polymer with a self-healing function.

Wang et al. [34] prepared a three-arm chain segment based on an effective host–guest inclusion interaction between ethyl-acrylate-modified β-cyclodextrin (β-CD-AOI2) and acryloyl-tetraethylene-glycol-modified adamantane (A-TEG-Ad). A host–guest supramolecular hydrogel (HGGelMA) was obtained by the reaction of double bonds at the end of the arms. The biocompatible natural matrix gelatin methacryloyl (GelMA) formed a covalent cross-linking network. The resulting hydrogel has non-covalent bonds embedded in the covalent connection network. Weak non-covalent bonds can be quickly re-established through host–guest recognition after cleavage, while strong covalent bonds maintain the network. The hydrogel has excellent biodegradability and low immune rejection of natural hydrogels as well as a high strength, anti-fatigue performance, and rapid healing rate. The shear-thinning behavior and suitable viscosity of the HGGelMA precursor meet the requirements for 3D printing. The hydrogel is a promising printing biomaterial with potential biomedical applications, circumventing some of the issues of traditional hydrogels.

The reversibility and repair processes of reversible covalent bonds require external energy stimulation, such as heating, light, and catalysts. For example, the temperature required for the self-healing of transesterification bonds and Diels–Alder bonds is usually higher than 100 °C, and the stimulation does not promote efficient healing in common environments. Although reversible non-covalent bonds can often achieve satisfactory healing effects under mild conditions, the mechanical performances are usually not as good as those formed by dynamic covalent interactions. Moreover, hydrogen bonds are usually repaired at room temperature, but their mechanical strength is limited. Both types of bonds have advantages and disadvantages.

4. Challenges and Prospects

Dynamic interactions endow UV-curable 3D-printed polymers with a self-healing function and recyclability, prolong the service life of products, and alleviate environmental pollution and resource shortages. The research field of dynamic UV-curable polymer 3D printing is gradually developing, and the accompanying difficulties and challenges must be solved urgently. Future research should address the following issues:

(1) Photocurable 3D printing technology requires photosensitive resin with a low viscosity, but the molecular weight of low-viscosity resin is small, which will make the cross-linking density of the cured material high, causing the material to become hard and brittle. If the molecular weight of the resin is large, a large amount of monomer dilution is required, which will cause the resin to lose its original performance. The contradiction between resin viscosity and performance needs to be solved urgently;

(2) Balancing the mechanical performance and self-healing function of UV-curable self-healing polymers is the main goal. To achieve high mechanical performance, dynamic bonds with high bond energy are required, which decreases the self-healing efficiency of the polymer. At present, dual dynamic network structures [103–105] and multi-phase design [106,107] have been introduced into polymers and have achieved certain results, but new methods still need to be explored;

(3) UV-curable, self-healing polymers for 3D printing require external stimuli to activate damage healing; however, the stimulation intensity required for healing cannot be easily provided in practical applications. Developing polymers that can self-heal at room temperature or lower is more valuable for practical applications;

(4) Currently, the preparation of self-healing photosensitive resins is generally complicated and requires cumbersome steps. Therefore, simplification of the synthesis process, improvement of the yield, and the reduction of waste are necessary;

(5) Photocuring printing equipment is usually expensive and mainly used for printing small devices. Using dynamic interaction, it is undoubtedly convenient to construct large-size devices through module assembly.

5. Conclusions

The demand for practical applications has prompted the rapid development of UV-curing 3D printing technology. However, it is necessary to simultaneously introduce self-healing and other functions into the printing polymers to develop functional photosensitive resins. This paper describes photosensitive resins with dynamic bonding for 3D printing that have been studied in recent years and categorizes the types of dynamic bonding. The design, preparation, and application prospects of polymers are reviewed. The advantages and disadvantages of the printed products are summarized. It is hoped that this review will be of some help for the subsequent development of UV-curing, self-healing polymers with excellent properties.

Author Contributions: Conceptualization, W.L. and G.Z.; writing—original draft preparation, W.L. and Z.S.; writing—review and editing, W.L., H.R., X.W. and G.Z.; formal analysis, T.Z. and W.W.; supervision, L.X. and G.Z.; funding acquisition, T.Z., W.W., L.X. and G.Z. All authors have read and agreed to the published version of the manuscript.

Funding: The National Natural Science Foundation of China (22105103, 12102194, 22205060, 21905084); the Natural Science Foundation of Jiangsu Province (BK20200471).

Data Availability Statement: No new data were created.

Conflicts of Interest: The authors declare no conflict of interest.

References

1. Yan, J.; Jiang, M.; Chen, J.; Liang, C. Automobile Body Integration Manufacturing and Thickness Optimization for Stereo Lithography 3D Printing. *J. Beijing Univ. Technol.* **2017**, *43*, 551–556.
2. Mohanavel, V.; Ali, K.S.A.; Ranganathan, K.; Jeffrey, J.A.; Ravikumar, M.M. The roles and applications of additive manufacturing in the aerospace and automobile sector. In Proceedings of the 12th National Web Conference on Recent Advancements in Biomedical Engineering (NCRABE), Electr Network, Online, 3 December 2021; pp. 405–409.
3. Srinivasan, D.; Meignanamoorthy, M.; Ravichandran, M.; Mohanavel, V.; Alagarsamy, S.V.; Chanakyan, C.; Sakthivelu, S.; Karthick, A.; Prabhu, T.R.; Rajkumar, S. 3D Printing Manufacturing Techniques, Materials, and Applications: An Overview. *Adv. Mater. Sci. Eng.* **2021**, *2021*, 5756563. [CrossRef]
4. Moroni, L.; Boland, T.; Burdick, J.A.; De Maria, C.; Derby, B.; Forgacs, G.; Groll, J.; Li, Q.; Malda, J.; Mironov, V.A.; et al. Biofabrication: A Guide to Technology and Terminology. *Trends Biotechnol.* **2018**, *36*, 384–402. [CrossRef] [PubMed]
5. Ligon, S.C.; Liska, R.; Stampfl, J.; Gurr, M.; Muelhaupt, R. Polymers for 3D Printing and Customized Additive Manufacturing. *Chem. Rev.* **2017**, *117*, 10212–10290. [CrossRef] [PubMed]
6. Huang, J.; Qin, Q.; Wang, J. A Review of Stereolithography: Processes and Systems. *Processes* **2020**, *8*, 1138. [CrossRef]
7. Fang, H.-B.; Chen, J.-M. 3D printing based on digital light processing technology. *J. Beijing Univ. Technol.* **2015**, *41*, 1775–1782. [CrossRef]
8. Saitta, L.; Tosto, C.; Pergolizzi, E.; Patti, A.; Cicala, G. Liquid Crystal Display (LCD) Printing: A Novel System for Polymer Hybrids Printing. In Proceedings of the 4th International Conference on Progress on Polymers and Composites Products and Manufacturing Technologies (POLCOM), Electr Network, Bucharest, Romania, 26–28 November 2020.
9. Braga, R.R.; Ballester, R.Y.; Ferracane, J.L. Factors involved in the development of polymerization shrinkage stress in resin-composites: A systematic review. *Dent. Mater.* **2005**, *21*, 962–970. [CrossRef]
10. Hayashi, J.; Espigares, J.; Takagaki, T.; Shimada, Y.; Tagami, J.; Numata, T.; Chan, D.; Sadr, A. Real-time in-depth imaging of gap formation in bulk-fill resin composites. *Dent. Mater.* **2019**, *35*, 585–596. [CrossRef]
11. Zhang, M.; Jiang, S.; Gao, Y.; Nie, J.; Sun, F. Design of a disulfide bond-containing photoresist with extremely low volume shrinkage and excellent degradation ability for UV-nanoimprinting lithography. *Chem. Eng. J.* **2020**, *390*, 124625. [CrossRef]
12. Zhang, Y.; Ju, X.; Gao, Y.; Sun, F. Design of a polymerizable dithioaniline derivative with double functions of reducing volume shrinkage and initiating polymerization for LED photopolymerization. *Eur. Polym. J.* **2022**, *179*, 111534. [CrossRef]
13. Lee, Y.S.; Park, W. Current challenges and future directions for bacterial self-healing concrete. *Appl. Microbiol. Biotechnol.* **2018**, *102*, 3059–3070. [CrossRef] [PubMed]

14. Sovljanski, O.; Tomic, A.; Markov, S. Relationship between Bacterial Contribution and Self-Healing Effect of Cement-Based Materials. *Microorganisms* **2022**, *10*, 1399. [CrossRef] [PubMed]

15. Grabowski, B.; Tasan, C.C. Self-Healing Metals. In *Self-Healing Materials*; Advances in Polymer Science; Hager, M.D., VanDerZwaag, S., Schubert, U.S., Eds.; Springer: Cham, Switzerland, 2016; Volume 273, pp. 387–407.

16. Li, Y.; Zhang, L.; Zhang, J.; Wang, X.; Gu, C.; Xia, X.; Tu, J. Self-Healing Properties of Alkali Metals under "High-Energy Conditions" in Batteries. *Adv. Energy Mater.* **2021**, *11*, 2100470. [CrossRef]

17. Shen, Y.; Yang, T.; Niu, X.; Du, Y. Research progress in extrinsic self-healing polymer composites. *Fiber Reinf. Plast. Compos.* **2015**, 92–96+63.

18. Wang, J.; Jia, H.; Fang, E.; Jiang, J.; Jiang, Q. Progress in self-healing of polymer composites. *China Synth. Rubber Ind.* **2012**, *35*, 247–253.

19. Wang, X.-X.; Yao, S.; Zhou, C.-J.; Wu, J.-L. Application and potential future directions of self-healing polymers in dentistry. *Huaxi Kouqiang Yixue Zazhi West China J. Stomatol.* **2020**, *38*, 75–79. [CrossRef]

20. Ye, B.-h.; Meng, L.; Li, L.-h.; Li, N.; Li, Z.-w. Self-healing Hydrogels Based on Constitutional Dynamic Chemistry and Their Potential Biomedical Applications. *Acta Polym. Sin.* **2016**, *2*, 134–148.

21. Li, S.; Yan, X.; Ma, S.; Liu, M.; Liu, J. Progress in the Application of Self-Healing Polymers in Leather Coatings. *Polym. Mater. Sci. Eng.* **2022**, *38*, 166–173.

22. Zhang, M.; Zhang, B.; Li, X.; Jin, J.; Wang, J. Progress in Research of Self-Healing Superhydrophobic Coatings. *Polym. Mater. Sci. Eng.* **2022**, *38*, 168–175.

23. Huynh, T.-P.; Sonar, P.; Haick, H. Advanced Materials for Use in Soft Self-Healing Devices. *Adv. Mater.* **2017**, *29*, 1604973. [CrossRef]

24. Lin, W.; Wang, H.; Zhang, H.; Tang, J.; Wang, Y. Advance in extrinsic polymer self-healing composite. *New Chem. Mater.* **2020**, *48*, 50–53.

25. Wang, X.; Cheng, B.; Liang, D.; Jia, D. Progress of Intrinsic Self-Healing Polymer Materials. *Polym. Mater. Sci. Eng.* **2019**, *35*, 183–190. [CrossRef]

26. He, Z.; Zhang, W.; Ma, W.; Yu, H.; Zhao, Y. Research progress of microcapsule intelligent self-healing material. *New Chem. Mater.* **2018**, *46*, 25–29.

27. Zhang, H.; Li, M.; Feng, X.; Han, X.; Fan, Z. Review of the Latest Research on Extrinsic Microcapsule Self-healing Materials. *Eng. Plast. Appl.* **2019**, *47*, 134–137.

28. Shinde, V.V.; Celestine, A.-D.; Beckingham, L.E.; Beckingham, B.S. Stereolithography 3D Printing of Microcapsule Catalyst-Based Self-Healing Composites. *ACS Appl. Polym. Mater.* **2020**, *2*, 5048–5057. [CrossRef]

29. Sanders, P.; Young, A.J.; Qin, Y.; Fancey, K.S.; Reithofer, M.R.; Guillet-Nicolas, R.; Kleitz, F.; Pamme, N.; Chin, J.M. Stereolithographic 3D printing of extrinsically self-healing composites. *Sci. Rep.* **2019**, *9*, 388. [CrossRef]

30. Roy, N.; Bruchmann, B.; Lehn, J.-M. DYNAMERS: Dynamic polymers as self-healing materials. *Chem. Soc. Rev.* **2015**, *44*, 3786–3807. [CrossRef] [PubMed]

31. White, S.R.; Sottos, N.R.; Geubelle, P.H.; Moore, J.S.; Kessler, M.R.; Sriram, S.R.; Brown, E.N.; Viswanathan, S. Autonomic healing of polymer composites. *Nature* **2001**, *409*, 794–797. [CrossRef]

32. Huang, W.; Zhang, J.; Singh, V.; Xu, L.; Kabi, P.; Bele, E.; Tiwari, M.K. Digital light 3D printing of a polymer composite featuring robustness, self-healing, recyclability and tailorable mechanical properties. *Addit. Manuf.* **2023**, *61*, 103343. [CrossRef]

33. Zhu, G.; Zhang, J.; Huang, J.; Yu, X.; Cheng, J.; Shang, Q.; Hu, Y.; Liu, C.; Zhang, M.; Hu, L.; et al. Self-Healing, Antibacterial, and 3D-Printable Polymerizable Deep Eutectic Solvents Derived from Tannic Acid. *ACS Sustain. Chem. Eng.* **2022**, *10*, 7954–7964. [CrossRef]

34. Wang, Z.; An, G.; Zhu, Y.; Liu, X.; Chen, Y.; Wu, H.; Wang, Y.; Shi, X.; Mao, C. 3D-printable self-healing and mechanically reinforced hydrogels with host-guest non-covalent interactions integrated into covalently linked networks. *Mater. Horiz.* **2019**, *6*, 733–742. [CrossRef] [PubMed]

35. Durand-Silva, A.; Cortés-Guzmán, K.P.; Johnson, R.M.; Perera, S.D.; Diwakara, S.D.; Smaldone, R.A. Balancing Self-Healing and Shape Stability in Dynamic Covalent Photoresins for Stereolithography 3D Printing. *ACS Macro Lett.* **2021**, *10*, 486–491. [CrossRef] [PubMed]

36. Wang, L.; Wang, X. Research Progress in Self-Healing Materials Based on Diels-Alder Reaction. *J. Funct. Polym.* **2014**, *27*, 453–463.

37. Grauzeliene, S.; Kazlauskaite, B.; Skliutas, E.; Malinauskas, M.; Ostrauskaite, J. Photocuring and digital light processing 3D printing of vitrimer composed of 2-hydroxy-2-phenoxypropyl acrylate and acrylated epoxidized soybean oil. *Express Polym. Lett.* **2023**, *17*, 54–68. [CrossRef]

38. Zhang, M.; Tao, X.; Yu, R.; He, Y.; Li, X.; Chen, X.; Huang, W. Self-healing, mechanically robust, 3D printable ionogel for highly sensitive and long-term reliable ionotronics. *J. Mater. Chem. A* **2022**, *10*, 12005–12015. [CrossRef]

39. Cortés-Guzmán, K.P.; Parikh, A.R.; Sparacin, M.L.; Remy, A.K.; Adegoke, L.; Chitrakar, C.; Ecker, M.; Voit, W.E.; Smaldone, R.A. Recyclable, Biobased Photoresins for 3D Printing Through Dynamic Imine Exchange. *ACS Sustain. Chem. Eng.* **2022**, *10*, 13091–13099. [CrossRef]

40. Zhang, Z.P.; Rong, M.Z.; Zhang, M.Q. Polymer engineering based on reversible covalent chemistry: A promising innovative pathway towards new materials and new functionalities. *Prog. Polym. Sci.* **2018**, *80*, 39–93. [CrossRef]

41. Chakma, P.; Konkolewicz, D. Dynamic Covalent Bonds in Polymeric Materials. *Angew. Chem. Int. Ed.* **2019**, *58*, 9682–9695. [CrossRef]
42. Billiet, T.; Vandenhaute, M.; Schelfhout, J.; Van Vlierberghe, S.; Dubruel, P. A review of trends and limitations in hydrogel-rapid prototyping for tissue engineering. *Biomaterials* **2012**, *33*, 6020–6041. [CrossRef]
43. Xenikakis, I.; Tsongas, K.; Tzimtzimis, E.K.; Zacharis, C.K.; Theodoroula, N.; Kalogianni, E.P.; Demiri, E.; Vizirianakis, I.S.; Tzetzis, D.; Fatouros, D.G. Fabrication of hollow microneedles using liquid crystal display (LCD) vat polymerization 3D printing technology for transdermal macromolecular delivery. *Int. J. Pharm.* **2021**, *597*, 120303. [CrossRef]
44. Tumbleston, J.R.; Shirvanyants, D.; Ermoshkin, N.; Janusziewicz, R.; Johnson, A.R.; Kelly, D.; Chen, K.; Pinschmidt, R.; Rolland, J.P.; Ermoshkin, A.; et al. Continuous liquid interface production of 3D objects. *Science* **2015**, *347*, 1349–1352. [CrossRef] [PubMed]
45. Quan, H.; Zhang, T.; Xu, H.; Luo, S.; Nie, J.; Zhu, X. Photo-curing 3D printing technique and its challenges. *Bioact. Mater.* **2020**, *5*, 110–115. [CrossRef] [PubMed]
46. Wang, S.; Zhu, Y.; Wu, Y.; Xiang, H.; Liu, X. Development and Applications of UV-Curing 3D Printing and Photosensitive Resin. *J. Funct. Polym.* **2022**, *35*, 19–35.
47. Šercer, M.; Godec, D.; Šantek, B.; Ludwig, R.; Andlar, M.; Rezić, I.; Ivušić, F.; Pilipović, A.; Oros, D.; Rezić, T. Microreactor Production by PolyJet Matrix 3D-Printing Technology: Hydrodynamic Characterization. *Food Technol. Biotechnol.* **2019**, *57*, 272–281. [CrossRef] [PubMed]
48. Yang, Y.-M.; Yu, H.-Z.; Sun, X.-H.; Dang, Z.-M. Density functional theory calculations on S-S bond dissociation energies of disulfides. *J. Phys. Org. Chem.* **2016**, *29*, 6–13. [CrossRef]
49. Rekondo, A.; Martin, R.; de Luzuriaga, A.R.; Cabañero, G.; Grande, H.J.; Odriozola, I. Catalyst-free room-temperature self-healing elastomers based on aromatic disulfide metathesis. *Mater. Horiz.* **2014**, *1*, 237–240. [CrossRef]
50. Nevejans, S.; Ballard, N.; Miranda, J.I.; Reck, B.; Asua, J.M. The underlying mechanisms for self-healing of poly(disulfide)s. *Phys. Chem. Chem. Phys.* **2016**, *18*, 27577–27583. [CrossRef]
51. Li, X.; Yu, R.; He, Y.; Zhang, Y.; Yang, X.; Zhao, X.; Huang, W. Self-Healing Polyurethane Elastomers Based on a Disulfide Bond by Digital Light Processing 3D Printing. *ACS Macro Lett.* **2019**, *8*, 1511–1516. [CrossRef]
52. Decker, C.; Viet, T.N.T. Photocrosslinking of functionalized rubbers, 8. The thiol-polybutadiene system. *Macromol. Chem. Phys.* **1999**, *200*, 1965–1974. [CrossRef]
53. Yu, K.; Xin, A.; Du, H.; Li, Y.; Wang, Q. Additive manufacturing of self-healing elastomers. *NPG Asia Mater.* **2019**, *11*, 7. [CrossRef]
54. Xu, Y.; Xiong, X.; Cai, L.; Tang, Z.; Ye, Z. Thiol-Ene Click Chemistry. *Prog. Chem.* **2012**, *24*, 385–394.
55. Yang, Z.; Chen, Q.; Zhou, D.; Bu, Y. Synthesis of Functional Polymer Materials via Thiol-Ene/Yne Click Chemistry. *Prog. Chem.* **2012**, *24*, 395–404.
56. Rahman, S.S.; Arshad, M.; Qureshi, A.; Ullah, A. Fabrication of a Self-Healing, 3D Printable, and Reprocessable Biobased Elastomer. *ACS Appl. Mater. Interfaces* **2020**, *12*, 51927–51939. [CrossRef]
57. Rattanangkool, E.; Krailat, W.; Vilaivan, T.; Phuwapraisirisan, P.; Sukwattanasinitt, M.; Wacharasindhu, S. Hypervalent Iodine(III)-Promoted Metal-Free S-H Activation: An Approach for the Construction of S-S, S-N, and S-C Bonds. *Eur. J. Org. Chem.* **2014**, *2014*, 4795–4804. [CrossRef]
58. Sijbesma, R.P.; Beijer, F.H.; Brunsveld, L.; Folmer, B.J.B.; Hirschberg, J.H.K.K.; Lange, R.F.M.; Lowe, J.K.L.; Meijer, E.W. Reversible polymers formed from self-complementary monomers using quadruple hydrogen bonding. *Science* **1997**, *278*, 1601–1604. [CrossRef]
59. Cordier, P.; Tournilhac, F.; Soulie-Ziakovic, C.; Leibler, L. Self-healing and thermoreversible rubber from supramolecular assembly. *Nature* **2008**, *451*, 977–980. [CrossRef] [PubMed]
60. Li, C.-H.; Wang, C.; Keplinger, C.; Zuo, J.-L.; Jin, L.; Sun, Y.; Zheng, P.; Cao, Y.; Lissel, F.; Linder, C.; et al. A highly stretchable autonomous self-healing elastomer. *Nat. Chem.* **2016**, *8*, 619–625. [CrossRef] [PubMed]
61. Gomez, E.F.; Wanasinghe, S.V.; Flynn, A.E.; Dodo, O.J.; Sparks, J.L.; Baldwin, L.A.; Tabor, C.E.; Durstock, M.F.; Konkolewicz, D.; Thrasher, C.J. 3D-Printed Self-Healing Elastomers for Modular Soft Robotics. *ACS Appl. Mater. Interfaces* **2021**, *13*, 28870–28877. [CrossRef] [PubMed]
62. Tong, X.; Feng, H.; Yan, Z.; Hao, Q.; Wang, L. Preparation and properties of carbon fiber reinforced unsaturated polyester self-healing composites. *Mod. Chem. Ind.* **2017**, *37*, 120–123,125.
63. Lei, R.; Ma, Y.; Yang, X. Self-healing performance of epoxy coating containing microencapsulated alkyd resin. *Chem. Ind. Eng. Prog.* **2020**, *39*, 2782–2787.
64. Huang, J.; Zhang, J.; Zhu, G.; Yu, X.; Hu, Y.; Shang, Q.; Chen, J.; Hu, L.; Zhou, Y.; Liu, C. Self-healing, high-performance, and high-biobased-content UV-curable coatings derived from rubber seed oil and itaconic acid. *Prog. Org. Coat.* **2021**, *159*, 106391. [CrossRef]
65. Grauzeliene, S.; Kastanauskas, M.; Talacka, V.; Ostrauskaite, J. Photocurable Glycerol- and Vanillin-Based Resins for the Synthesis of Vitrimers. *ACS Appl. Polym. Mater.* **2022**, *4*, 6103–6110. [CrossRef] [PubMed]
66. Yuan, Y.; Yang, J.; Wu, Q.; Wu, M.; Zhang, J.a. Preparation and properties of room temperature self-healing waterborne polyurethane based on imine bond and disulfide bond. *Fine Chem.* **2022**, *39*, 2449–2455,2466.

67. Pan, Z.; Zhao, Q.; Xue, Y.; Bo, C.; Zhang, M. Preparation and Properties of Bio-Based Self-Healing and Recyclable Polyurethane Elastomer Based on Dynamic Iimine Bonds. *Polym. Mater. Sci. Eng.* **2022**, *38*, 11–18.
68. Min, J.; Zhou, Z.; Wang, H.; Chen, Q.; Hong, M.; Fu, H. Room temperature self-healing and recyclable conductive composites for flexible electronic devices based on imine reversible covalent bond. *J. Alloys Compd.* **2022**, *894*, 162433. [CrossRef]
69. Liguori, A.; Subramaniyan, S.; Yao, J.G.; Hakkarainen, M. Photocurable extended vanillin-based resin for mechanically and chemically recyclable, self-healable and digital light processing 3D printable thermosets. *Eur. Polym. J.* **2022**, *178*, 111489. [CrossRef]
70. Miao, J.-T.; Ge, M.; Peng, S.; Zhong, J.; Li, Y.; Weng, Z.; Wu, L.; Zheng, L. Dynamic Imine Bond-Based Shape Memory Polymers with Permanent Shape Reconfigurability for 4D Printing. *ACS Appl. Mater. Interfaces* **2019**, *11*, 40642–40651. [CrossRef]
71. Fache, M.; Boutevin, B.; Caillol, S. Vanillin Production from Lignin and Its Use as a Renewable Chemical. *ACS Sustain. Chem. Eng.* **2016**, *4*, 35–46. [CrossRef]
72. Chen, X.; Dam, M.A.; Ono, K.; Mal, A.; Shen, H.; Nutt, S.R.; Sheran, K.; Wudl, F. A thermally re-mendable cross-linked polymeric material. *Science* **2002**, *295*, 1698–1702. [CrossRef]
73. Turkenburg, D.H.; Fischer, H.R. Diels-Alder based, thermo-reversible cross-linked epoxies for use in self-healing composites. *Polymer* **2015**, *79*, 187–194. [CrossRef]
74. Liu, Y.-L.; Chuo, T.-W. Self-healing polymers based on thermally reversible Diels-Alder chemistry. *Polym. Chem.* **2013**, *4*, 2194–2205. [CrossRef]
75. Li, Q.-T.; Jiang, M.-J.; Wu, G.; Chen, L.; Chen, S.-C.; Cao, Y.-X.; Wang, Y.-Z. Photothermal Conversion Triggered Precisely Targeted Healing of Epoxy Resin Based on Thermoreversible Diels-Alder Network and Amino-Functionalized Carbon Nanotubes. *ACS Appl. Mater. Interfaces* **2017**, *9*, 20797–20807. [CrossRef] [PubMed]
76. Turkenburg, D.H.; Durant, Y.; Fischer, H.R. Bio-based self-healing coatings based on thermo-reversible Diels-Alder reaction. *Prog. Org. Coat.* **2017**, *111*, 38–46. [CrossRef]
77. Liu, M.; Zhong, J.; Li, Z.; Rong, J.; Yang, K.; Zhou, J.; Shen, L.; Gao, F.; Huang, X.; He, H. A high stiffness and self-healable polyurethane based on disulfide bonds and hydrogen bonding. *Eur. Polym. J.* **2020**, *124*, 109475. [CrossRef]
78. Herbst, F.; Doehler, D.; Michael, P.; Binder, W.H. Self-Healing Polymers via Supramolecular Forces. *Macromol. Rapid Commun.* **2013**, *34*, 203–220. [CrossRef]
79. Ma, S.; Qi, X.; Cao, Y.; Yang, S.; Xu, J. Hydrogen bond detachment in polymer complexes. *Polymer* **2013**, *54*, 5382–5390. [CrossRef]
80. Wu, Y.; Fei, M.; Chen, T.; Li, C.; Wu, S.; Qiu, R.; Liu, W. Photocuring Three-Dimensional Printing of Thermoplastic Polymers Enabled by Hydrogen Bonds. *ACS Appl. Mater. Interfaces* **2021**, *13*, 22946–22954. [CrossRef]
81. Wu, Y.; Fei, M.; Chen, T.; Li, C.; Fu, T.; Qiu, R.; Liu, W. H-bonds and metal-ligand coordination-enabled manufacture of palm oil-based thermoplastic elastomers by photocuring 3D printing. *Addit. Manuf.* **2021**, *47*, 102268. [CrossRef]
82. Li, R.; Fan, T.; Chen, G.; Xie, H.; Su, B.; He, M. Highly transparent, self-healing conductive elastomers enabled by synergistic hydrogen bonding interactions. *Chem. Eng. J.* **2020**, *393*, 124685. [CrossRef]
83. Lai, C.-W.; Yu, S.-S. 3D Printable Strain Sensors from Deep Eutectic Solvents and Cellulose Nanocrystals. *ACS Appl. Mater. Interfaces* **2020**, *12*, 34235–34244. [CrossRef]
84. Cai, L.; Chen, G.; Su, B.; He, M. 3D printing of ultra-tough, self-healing transparent conductive elastomeric sensors. *Chem. Eng. J.* **2021**, *426*, 130545. [CrossRef]
85. Shirmohammadli, Y.; Efhamisisi, D.; Pizzi, A. Tannins as a sustainable raw material for green chemistry: A review. *Ind. Crops Prod.* **2018**, *126*, 316–332. [CrossRef]
86. Lim, B.-S.; Lee, Y.-K.; 신현철; 최재윤. The Effect of Weight fraction of Photo Initiator and Inhibitor in the Expreimental Resins on the Dgree of Polymerization. *Korean J. Dent. Mater.* **2004**, *31*, 299–306.
87. Bennett, J. Measuring UV curing parameters of commercial photopolymers used in additive manufacturing. *Addit. Manuf.* **2017**, *18*, 203–212. [CrossRef] [PubMed]
88. Li, Y.; Kankala, R.K.; Wu, L.; Chen, A.-Z.; Wang, S.-B. 3D-Printed Photocurable Resin with Synergistic Hydrogen Bonding Based on Deep Eutectic Solvent. *ACS Appl. Polym. Mater.* **2023**, *5*, 991–1001. [CrossRef]
89. Zhu, G.; Hou, Y.; Xiang, J.; Xu, J.; Zhao, N. Digital Light Processing 3D Printing of Healable and Recyclable Polymers with Tailorable Mechanical Properties. *ACS Appl. Mater. Interfaces* **2021**, *13*, 34954–34961. [CrossRef] [PubMed]
90. Wu, Q.; Han, S.; Zhu, J.; Chen, A.; Zhang, J.; Yan, Z.; Liu, J.; Huang, J.; Yang, X.; Guan, L. Stretchable and self-healing ionic conductive elastomer for multifunctional 3D printable sensor. *Chem. Eng. J.* **2023**, *454*, 140328. [CrossRef]
91. Invernizzi, M.; Turri, S.; Levi, M.; Suriano, R. Processability of 4D printable modified polycaprolactone with self-healing abilities. In Proceedings of the 1st International Conference on Materials, Mimicking, Manufacturing from and for Bio Application (BioM&M), Milan, Italy, 27–29 June 2019; pp. 508–515.
92. Suriano, R.; Bernasconi, R.; Magagnin, L.; Levi, M. 4D Printing of Smart Stimuli-Responsive Polymers. *J. Electrochem. Soc.* **2019**, *166*, B3274–B3281. [CrossRef]
93. Durand-Silva, A.; Perera, S.D.; Remy, A.K.; Peng, H.-C.; Hargrove, J.A.; Ferneyhough, Z.D.; Landaverde, P.M.L.; Stelling, A.L.; Smaldone, R.A. Infrared Spectroscopic and Mechanical Analysis of Supramolecular Self-Healing in 3D Printable Urea Photoresins. *ACS Appl. Polym. Mater.* **2022**, *4*, 8825–8832. [CrossRef]
94. Chen, X.; Zawaski, C.E.; Spiering, G.A.; Liu, B.; Orsino, C.M.; Moore, R.B.; Williams, C.B.; Long, T.E. Quadruple Hydrogen Bonding Supramolecular Elastomers for Melt Extrusion Additive Manufacturing. *ACS Appl. Mater. Interfaces* **2020**, *12*, 32006–32016. [CrossRef]

95. Wu, Y.; Zeng, Y.; Chen, Y.; Li, C.; Qiu, R.; Liu, W. Photocurable 3D Printing of High Toughness and Self-Healing Hydrogels for Customized Wearable Flexible Sensors. *Adv. Funct. Mater.* **2021**, *31*, 2107202. [CrossRef]
96. Liu, Y.; Wang, Y.; Yu, J.; Zhu, J.; Hu, Z. Effect on Performance of Reversible Covalent Self-healing Materials of Sacrificial Bonds. *J. Donghua Univ. Nat. Sci. Ed.* **2018**, *44*, 45–52.
97. Flow-Induced Crystallization in Polymer Systems. 1979, p. x+370. Available online: https://sci-hub.yt/10.1016/0022-2860(80)852 07-0 (accessed on 24 September 2023).
98. Movsisyan, K.A.; Gasparyan, R.A.; Ovsepyan, A.M. Crystallization kinetics of crosslinked polymers. *Sov. J. Contemp. Phys.* **1990**, *25*, 51–53.
99. Zhang, B.; Zhang, W.; Zhang, Z.; Zhang, Y.-F.; Hingorani, H.; Liu, Z.; Liu, J.; Ge, Q. Self-Healing Four-Dimensional Printing with an Ultraviolet Curable Double-Network Shape Memory Polymer System. *ACS Appl. Mater. Interfaces* **2019**, *11*, 10328–10336. [CrossRef] [PubMed]
100. Wen, J.; Chen, T.; Wang, J.; Tuo, X.; Gong, Y.; Guo, J. Study on the healing performance of poly(epsilon-caprolactone) filled ultraviolet-curable 3D printed cyclic trimethylolpropane formal acrylate shape memory polymers. *J. Appl. Polym. Sci.* **2022**, *139*, e53085. [CrossRef]
101. Abdullah, T.; Okay, O. 4D Printing of Body Temperature-Responsive Hydrogels Based on Poly(acrylic acid) with Shape-Memory and Self-Healing Abilities. *ACS Appl. Bio Mater.* **2023**, *6*, 703–711. [CrossRef]
102. Chen, G.; Jiang, M. Cyclodextrin-based inclusion complexation bridging supramolecular chemistry and macromolecular self-assembly. *Chem. Soc. Rev.* **2011**, *40*, 2254–2266. [CrossRef]
103. An, S.Y.; Noh, S.M.; Oh, J.K. Multiblock Copolymer-Based Dual Dynamic Disulfide and Supramolecular Crosslinked Self-Healing Networks. *Macromol. Rapid Commun.* **2017**, *38*, 1600777. [CrossRef]
104. Cheng, B.; Lu, X.; Zhou, J.; Qin, R.; Yang, Y. Dual Cross-Linked Self-Healing and Recyclable Epoxidized Natural Rubber Based on Multiple Reversible Effects. *ACS Sustain. Chem. Eng.* **2019**, *7*, 4443–4455. [CrossRef]
105. Zhang, Q.; Niu, S.; Wang, L.; Lopez, J.; Chen, S.; Cai, Y.; Du, R.; Liu, Y.; Lai, J.C.; Liu, L.; et al. An Elastic Autonomous Self-Healing Capacitive Sensor Based on a Dynamic Dual Crosslinked Chemical System. *Adv. Mater.* **2018**, *30*, 1801435. [CrossRef]
106. Tolvanen, J.; Nelo, M.; Alasmäki, H.; Siponkoski, T.; Mäkelä, P.; Vahera, T.; Hannu, J.; Juuti, J.; Jantunen, H. Ultraelastic and High-Conductivity Multiphase Conductor with Universally Autonomous Self-Healing. *Adv. Sci.* **2022**, *9*, 2205485. [CrossRef] [PubMed]
107. Xu, J.; Chen, P.; Wu, J.; Hu, P.; Fu, Y.; Jiang, W.; Fu, J. Notch-Insensitive, Ultrastretchable, Efficient Self-Healing Supramolecular Polymers Constructed from Multiphase Active Hydrogen Bonds for Electronic Applications. *Chem. Mater.* **2019**, *31*, 7951–7961. [CrossRef]

Article

Exploring Bio-Based Polyurethane Adhesives for Eco-Friendly Structural Applications: An Experimental and Numerical Study

Ana M. S. Couto [1], Catarina S. P. Borges [2], Shahin Jalali [2], Beatriz D. Simões [2], Eduardo A. S. Marques [1], Ricardo J. C. Carbas [2,*], João C. Bordado [3], Till Vallée [4] and Lucas F. M. da Silva [1]

[1] Departamento de Engenharia Mecânica, Faculdade de Engenharia, Universidade do Porto, Rua Dr. Roberto Frias, 4200-465 Porto, Portugal
[2] Institute of Science and Innovation in Mechanical and Industrial Engineering (INEGI), Rua Dr. Roberto Frias 400, 4200-465 Porto, Portugal
[3] Centro de Recursos Naturais e Ambiente (CERENA), Instituto Superior Técnico, University of Lisbon, 1049-001 Lisbon, Portugal
[4] Fraunhofer Institute for Manufacturing Technology and Advanced Materials (IFAM), Wiener Str. 12, 28359 Bremen, Germany
* Correspondence: rcarbas@fe.up.pt

Abstract: In response to heightened environmental awareness, various industries, including the civil and automotive sector, are contemplating a shift towards the utilization of more sustainable materials. For adhesive bonding, this necessitates the exploration of materials derived from renewable sources, commonly denoted as bio-adhesives. This study focuses on a bio-adhesive L-joint, which is a commonly employed configuration in the automotive sector for creating bonded structural components with significant bending stiffness. In this investigation, the behavior of joints composed of pine wood and bio-based adhesives was studied. Two distinct configurations were studied, differing solely in the fiber orientation of the wood. The research combined experimental testing and finite element modeling to analyze the strength of the joints and determine their failure mode when subjected to tensile loading conditions. The findings indicate that the configuration of the joint plays a crucial role in its overall performance, with one of the solutions demonstrating higher strength. Additionally, a good degree of agreement was observed between the experimental and numerical analyses for one of the configurations, while the consideration of the maximum principal stress failure predictor (MPSFP) proved to accurately predict the location for crack propagation in both configurations.

Keywords: automotive industry; adhesive bonding; L-joint; finite element analysis; pine wood; biomaterials

Citation: Couto, A.M.S.; Borges, C.S.P.; Jalali, S.; Simões, B.D.; Marques, E.A.S.; Carbas, R.J.C.; Bordado, J.C.; Vallée, T.; da Silva, L.F.M. Exploring Bio-Based Polyurethane Adhesives for Eco-Friendly Structural Applications: An Experimental and Numerical Study. *Polymers* **2024**, *16*, 2546. https://doi.org/10.3390/polym16172546

Academic Editors: Guangpu Zhang, Zhengmao Ding and Yanan Zhang

Received: 29 June 2024
Revised: 26 August 2024
Accepted: 3 September 2024
Published: 9 September 2024

1. Introduction

In recent times, there has been a concerted effort to reduce the impact of fossil-based products, leading to the development and adoption of bio-based materials [1–3]. This movement aims to create a more sustainable future by emphasizing the use of tree- and plant-derived products. Specifically, in the construction of eco-friendly load-bearing structures, these materials show great promise, allowing for the creation of composites with natural fibers like flax, jute, and palm. Wood, a timeless resource, has played a significant role throughout history due to its renewable nature [4,5]. Wood, with its multifaceted qualities, stands out as an enticing material choice. Its versatility allows it to take on various shapes, while its impressive durability ensures resilience against wear and environmental factors. The mechanical robustness of wood, combined with its relatively low weight, makes it an appealing option for load-bearing structures. Moreover, its global abundance and economic viability further enhance its allure. What truly sets wood apart is its adaptability. It responds to changing environmental conditions, including temperature fluctuations, varying loading rates, and moisture influences. This adaptability has led to the creation of intricate artifacts with refined geometries a testament to wood's timeless appeal [6,7]. Wood, as a natural composite material, shares similarities with other composites. However, it does exhibit sensitivity when holes and notches are introduced. Interestingly, traditional joining methods like riveting and bolting are ill suited for wood due to its unique composition [8].

In structural applications, wooden adhesive joints play a pivotal role across diverse industries. In civil engineering, adhesive bonding provides uniform stress distribution without the need for local heating or substrate modifications [2,9]. These joints find use in bridges, buildings, and other infrastructure, enhancing structural integrity and enabling innovative designs. Similarly, in the automotive industry, wood-based adhesive joints offer lightweight and flexible connections, reducing the reliance on mechanical fasteners. Applications include vehicle interiors, panels, and non-structural components. Overall, wooden adhesive joints contribute to sustainable and efficient designs, making them valuable for the future. Bio-based adhesives offer several advantages over traditional petroleum-based counterparts, particularly in terms of sustainability and environmental impact reduction. Derived from renewable resources, these adhesives contribute to a lower carbon footprint by emitting fewer greenhouse gases during production. Additionally, they exhibit lower toxicity to both humans and the environment. In wooden applications, bio-based polyurethane adhesives can be modified for enhanced water resistance and bonding strength [5,10]. These innovative materials pave the way for high-performance alternatives, matching or surpassing synthetic wood adhesives while promoting eco-conscious practices. Adhesively bonded joints offer several advantages over traditional mechanical fasteners (such as bolts, rivets, and welds). First, they distribute stress more uniformly, minimizing stress concentrations. Second, adhesive bonding enhances shock and impact resistance, making joints more robust. Third, adhesives allow effective joining of dissimilar materials, including metals, plastics, wood, ceramics, and more. Fourth, adhesive bonds reduce weight by eliminating the need for additional hardware. Finally, these joints create seamless connections without visible cuts or holes, providing a cleaner overall appearance [11].

In the context of eco-friendly adhesive bonding for wooden structures in industrial applications, various joint configurations are commonly employed. Bio-based adhesives, derived from renewable resources, offer environmental advantages. Among the joint designs, Single-Lap Joints (SLJ) involve overlapping two wooden pieces with adhesive, while Double-Lap Joints (DLJ) provide better load distribution and increased strength. Stepped-Lap Joints distribute stress evenly due to their stepped profile, and Scarf Joints, with tapered overlaps, reduce stress concentration. Additionally, T-joints, where three wooden members intersect, play a crucial role in structural connections. However, the focus lies on L-joints, which are formed when two members meet at a right angle. L-joints are prevalent in wooden frames, furniture, and cabinetry, and their adhesive performance significantly impacts overall structural integrity. The choice of adhesive should consider material properties, load conditions, and environmental factors [12–14].

When addressing the numerical simulation of wooden joints, particularly within the context of structural applications, several critical computational aspects come into play. Wood, as a heterogeneous and highly anisotropic material, exhibits ductile behavior under compression but tends toward brittle behavior when subjected to tension or shear forces. To tackle this inherent complexity, researchers have developed advanced 3D constitutive models based on continuum damage mechanics. These models allow for precise representation of wood behavior and have been successfully integrated into finite element frameworks. The validation process, involving embedment and joint tests across different wood species (such as spruce, beech, and azobé), confirms the accuracy of these models by identifying failure modes that align with experimental observations [15–17].

Numerical simulations have become indispensable tools in product design and development across various industries, offering substantial cost savings and efficiency improvements [18,19] However, it is worth noting that engineering practices leveraging numerical simulations are not yet widely adopted throughout the entire forest wood chain. Currently, three main approaches characterize the mathematical description of wood material: (1) employing homogeneous volume elements that simplify wood properties and structure [6,20–22]; (2) simulating the wood cell wall and structure at a micro-level [23]; and (3) utilizing multiscale modeling techniques [24–26]. While numerical methods have found extensive application in civil engineering, particularly in studying deformations under static loads, material failure, and crack propagation using finite element methods (FEM) [27–29], the material failure of wood remains an area that has not been thoroughly explored using numerical simulations [30,31]. In the specific case of adhesive joints, cohesive zone elements strike a reasonable balance between calculation time and prediction accuracy.

Numerical models require prior knowledge of material properties. In the context of adhesive joints with biomaterials, the material properties of the substrates and adhesives must be carefully evaluated. Therefore, an experimental investigation must be conducted to characterize these materials. Once the properties are successfully obtained, the numerical model can simulate the overall behavior of the joint. Additional experimental studies within the overall joint must be performed, for validation of the model. Moura and Dourado [17], conducted a study that presented various practical applications with pine wood. These applications encompassed the repair of beams under tensile loading, repair of beams under bending loading, reinforcement of wood structures, steel–wood–steel connections, and wood–wood joints. The researchers utilized typical wood connections and structural details to emphasize the importance of incorporating non-linear fracture mechanics concepts, particularly cohesive zone modeling, in this field. The study achieved a comprehensive qualitative and quantitative representation of the mechanical behavior and failure modes observed in these applications. Hence, its work establishes the proposed procedures as fundamental tools for the efficient design of wood structures. In a separate study, Corte-Real et al., in 2022 [32], investigated single lap joints (SLJs) using different types of wooden substrates. These substrates included natural wood, wood/cork composites, and densified pine wood bonded with a novel polyurethane-based bio-adhesive. It was concluded that the densification process was successful, although it presented an additional challenge due to the resulting surface. Furthermore, increasing the overlap had a positive impact on the energy absorption of the joint, and the addition of cork resulted in a more consistent failure mode and higher strain to failure.

This study focuses on pine wood, a material that is both economically viable and abundantly available. Despite prior research on pine wood, there has been limited exploration of its suitability in L-joint configurations, which are commonly employed in structural applications within the civil and automotive industries. To address this gap, extensive research is essential to comprehend the behavior of adhesively bonded wooden L-joints.

The objective of this research work is to provide design tools for constructing more complex structures using wood. Novel wood-based materials, such as densified wood, hold promise for eventually replacing metals and composites while enhancing sustainability. In this study, the potential of pine wood in L-joint configurations subjected to tension loading conditions was investigated. To achieve this goal, two different joint configurations conducted, an experimental study involving. Additionally, a numerical analysis complemented these experimental efforts.

The findings from this research contribute valuable insights for designing load-bearing structures using pine wood, paving the way for eco-friendly alternatives in various industries.

2. Materials and Methods

In this section, a careful description of the materials used, including their properties and geometry and dimensions of the joint are provided. Additionally, the testing conditions of all experiments are revealed to ensure transparency and reproducibility of the results. Finally, this section presents the conclusive findings obtained from the experiments.

2.1. Materials

2.1.1. Substrate

In this study, pine wood served as the primary substrate, sourced from the southern regions of Portugal. The selection was based on its cost-effectiveness, widespread availability, and suitable mechanical properties. Notably, the mechanical behavior of wood is intricate and highly dependent on various factors, including species, growth conditions, grain slope and size, defects, knots, shakes, and age. For bio-substrate applications, beams exhibiting maximum symmetry and straight grains were chosen, while the use of knotted timber was avoided. The mechanical and cohesive properties of pine wood were experimentally determined by Moura et al. [21] and are presented in Tables 1–3.

Table 1. Nominal elastic properties of Pinus pinaster. (Young's modulus (E), Poisson's ratio (v), and shear modulus (G)) [17].

E_L [GPa]	E_R [GPa]	E_T [GPa]	v_{TL}	v_{RL}	v_{RT}	G_{RL} [GPa]	G_{TL} [GPa]	G_{RT} [GPa]
12.0	1.91	1.01	0.51	0.47	0.31	1.12	1.04	0.29

Table 2. Nominal strength properties of Pinus pinaster [17].

σ_{uL} [MPa]	σ_{uR} [MPa]	σ_{uT} [MPa]	σ_{uRL} [MPa]	σ_{uTL} [MPa]	σ_{uRT} [MPa]
97.5	7.90	4.20	16.0	16.0	4.50

Table 3. Cohesive parameters of Pinus pinaster in the RL fracture system [17].

G_{Ic} [N/mm]	G_{IIc} [N/mm]	$\sigma_{1,I}$ [MPa]	$\sigma_{1,II}$ [MPa]
0.264	0.94	5.34	9.27

The strength and fracture characteristics of wood significantly depend on the direction of loading. Neglecting this consideration could lead to severe wood failure. Wood cells align with the grain direction, resulting in maximum strength when the load is applied along this axis. To describe wood's mechanical and physical properties, a cylindrical coordinate system—comprising longitudinal (L), radial (R), and tangential (T) directions—is commonly employed due to the circular grain orientation (see Figure 1).

The moisture content of the wood fell within the typical range of 8% to 25% by weight, suitable for various wood types and applications. It remained dry, with a moisture content not exceeding 19%, which aligns with the limit for sawn lumber design. This moisture level significantly impacts the wood's properties, including dimensional stability, strength, durability, and resistance to biological factors.

For a better understanding of the different planes considered in this study, Figure 1 presents the main orthotropic directions and planes in wood.

2.1.2. Bio-Adhesive

A bio-based moisture-curing polyurethane, composed of 70% natural materials, was employed to bond the substrates. As a moisture-curing adhesive, it reacts with the moisture present in the substrate or the surrounding air during the curing process. To achieve a flawless bond without voids or defects, the adhesive must directly contact the substrates in a zero-thickness condition, allowing it to interact with the hydroxyl (OH) groups in the wood. This interaction exploits both mechanical interlocking and chemical bonding with the wooden substrates. These chemical bonds form between the OH groups in the wood and key components of the adhesive, synergistically enhancing the overall adhesive strength. Notably, the moisture content of the wood significantly influences the curing process of this bio-adhesive. Maintaining a consistent moisture level is crucial for determining the adhesive's curing time. Therefore, all substrates used in this study adhered to a specified moisture content range of 12–20%, meticulously ensured by the wood supplier. This precise control of moisture content ensures uniformity in the adhesive's curing process, ultimately enhancing its performance and reliability. The adhesive cures at 100 °C for a duration of 8 h. It is essential to mention that this bio-adhesive is currently in the prototype stage and has not been commercially released; however, it shows potential as a sustainable alternative to synthetic adhesives. It is produced in an irreversible reaction, without humidity, in a reactor under a nitrogen atmosphere and heating is performed with a thermal oil coil. It uses an aliphatic isocyanate as a basis, which contains 70% plant matter, and thus is more easily biodegradable. Manufacturing the bio-adhesive is estimated to consume 15 to 20% less energy than those derived from petroleum. The adhesive was developed by Professor João Bordado from Instituto Superior Técnico and was specifically formulated to display excellent compatibility and adhesion to wood substrates.

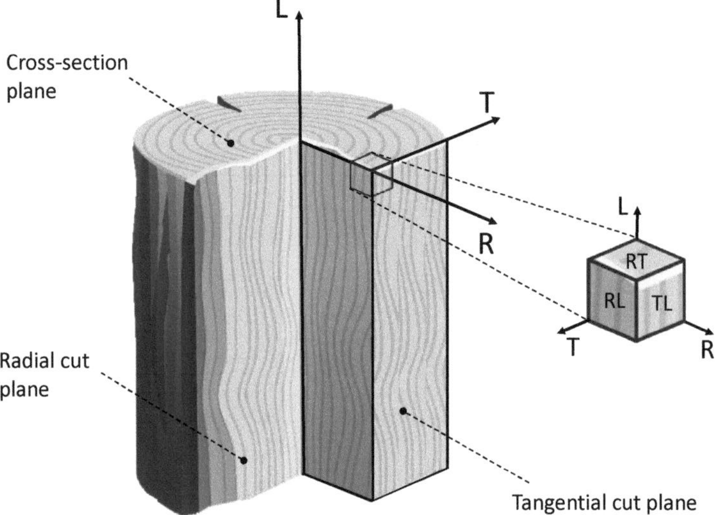

Figure 1. Main orthotropic directions and planes in wood.

The experimental characterization of the adhesive properties was conducted by Jalali et al. [33]. In terms of mechanical characterization, testing using a modified thick adherend shear test (TAST) specimen revealed a minimum shear strength value of 12.4 MPa. The author mentions tests utilizing modified double-cantilever beam (DCB) and end-notched flexure (ENF) specimens provided G_{Ic} and G_{IIc} values of 0.16 N/mm. In butt joints, the tensile strength was measured to be 16.5 MPa.

2.2. Joint Manufacturing

The joint assembly comprises an L-shaped upper substrate and a flat bottom substrate. When manufacturing joints using orthotropic wood, one critical parameter is the orientation of the wood rings. Depending on the substrate, various orientations were deliberately considered. For the bottom substrate, the preferred ring alignment is depicted in Figure 2. However, achieving flawless symmetry around a centrally aligned vertical axis was not consistently feasible. As for the upper substrate, two distinct configurations were studied: a nominally vertical orientation (a) and an inclined one (b). Although the term 'vertical' simplifies the description, it is important to note that the actual orientation may not be perfectly vertical. By observing both orientations in Figure 2, it can be understand the differences in upper substrate alignment.

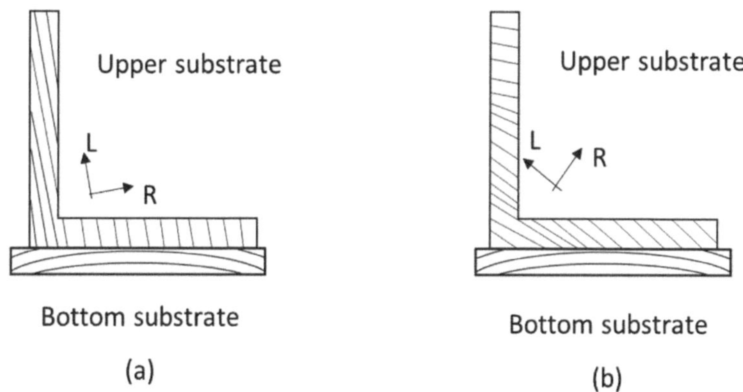

Figure 2. Illustration of the cross-section dimensions of the joints: (**a**) vertical orientation of the upper substrate and (**b**) inclined orientation of the upper substrate.

The manufacturing process involved preparing the substrates, which had previously been conditioned at 23 °C and 50% rH. To ensure a similar topography of the substrates, all specimens were cut in the same equipment and from wood extracted from the same batch. This was followed simply by air blasting to remove all dust from the surface. This process produced surfaces that were consistent in their surface roughness. A thin layer of adhesive was evenly spread across the overlap region on both substrates. The substrates were then carefully aligned and firmly pressed together. To ensure a robust bond, pressure was exerted on the overlap area using a plate and clamps. The assembled substrates were subsequently positioned in an oven to ensure the curing of the adhesive. Finally, the joints were removed, and any surplus adhesive along the edges was cleaned as needed. Figure 3 illustrates the final assembly and provides an overview of the joint dimensions. After the assembly, all samples were again conditioned at 23 °C and 50% rH.

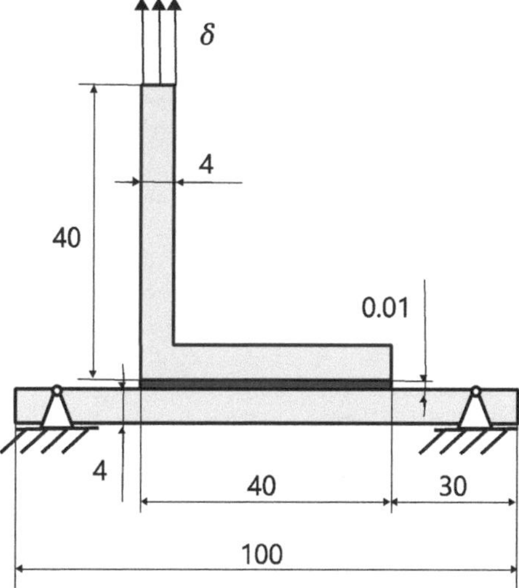

Figure 3. Dimensions of the L-joints, in mm, and the direction of the displacement, δ.

2.3. Testing Conditions

The L-joints manufactured for this work were tested via the application of a tensile load on the upper substrate and were conducted by the quasi-static test machine. The tensile tests were performed using a universal testing machine INSTRON 3367® (Illinois Tool Works, Hopkinton, MA, USA) with a load cell capacity of 30 kN, and both the load, *P*, and displacement, *δ*, were recorded by the machine at a constant rate of 1 mm/min. Figures 3 and 4 show the dimensions of the specimens and the setup utilized to conduct the testing procedures. Furthermore, the width of upper and bottom substrate was 15 and 25 mm, respectively. The clamped length was controlled during the tests, ensuring that it was always set to be 15 mm. It should be noted that at least five specimens were tested for each condition. The tests were conducted in laboratory conditions of temperature and humidity.

Figure 4. Experimental setup for testing the L-joints.

2.4. Numerical Details

A finite element model was developed to simulate the L-joints in the attempt to predict the real behavior of the structure. The Abaqus 2022® software was used to carry out the analysis. Numerical models for the two previously mentioned configurations (Figure 2) were made considering the mechanical properties previously reported for the pine wood and the bio-adhesive.

To model the adhesive layer, cohesive elements were utilized, and their behavior was defined by a triangular traction-separation law. The properties of this law, including strength and fracture properties of the novel bio-adhesive, were determined by Jalali et al. [33]. For the wood substrates, a homogenous orthotropic layer was employed with the corresponding elastic properties presented in Tables 1–3. Additionally, a layer of cohesive elements was positioned at a 0.15 mm distance of the interface of both upper and bottom substrates to allow for possible delamination of the wood. This distance was selected according to experimental observation, which has shown that this type of failure occurs one or two fibers away from the surface. This usually corresponds to a distance between 0.1 and 0.2 mm from the surface. The triangular cohesive law governing the elements in the wood was defined based on the strength and fracture properties of natural pine wood, as determined through experimental analysis by Moura and Dourado [17]. A visual representation of the model can be seen in Figure 5.

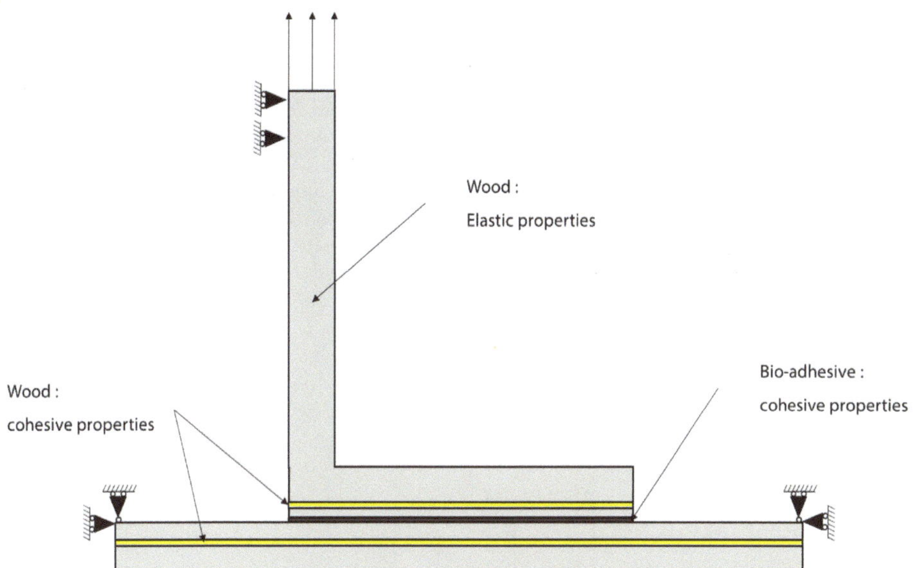

Figure 5. Boundary conditions and cohesive elements used in the numerical model of the L-joint.

In this study, we implemented boundary conditions and mesh configurations meticulously to ensure precise finite element simulations. First, we fixed the horizontal and vertical directions of the two upper edges of the bottom substrate. Simultaneously, the upper edge of the upper substrate experienced vertical upward displacement while remaining horizontally constrained. To achieve reliable results, we employed a high mesh density, comprising 11,740 four-node bilinear plane strain quadrilateral elements (CPE4) and 700 four-node two-dimensional cohesive elements (COH2D4), as depicted in Figure 6. Additionally, we discretized the substrate thickness using over 20 elements to capture bending phenomena accurately. Notably, a mesh refinement strategy was applied to account for localized stress effects in the adhesive region. These considerations collectively enhance the accuracy and reliability of our FEM simulations.

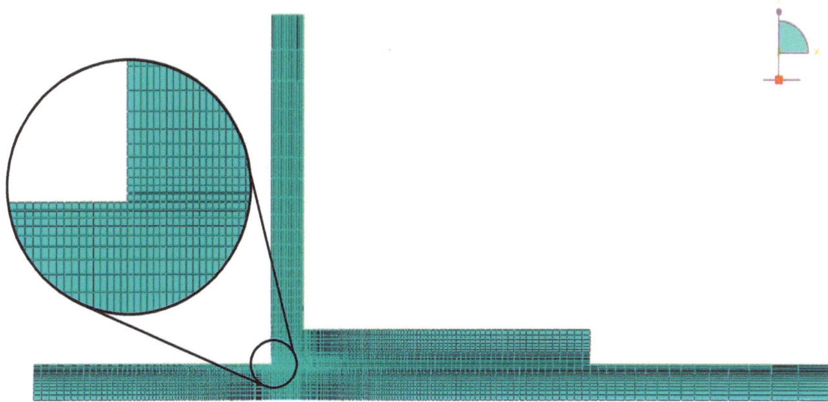

Figure 6. Mesh applied to the L-joint.

3. Experimental and Numerical Analysis

3.1. Strength Analysis

A strength analysis was conducted considering both configurations. The peel strength—displacement curves were derived from experimental procedures, directly extracted from the testing machine. The load retrieved was divided by the upper substrate width in order to determine the strength of each tested specimen. It is important to highlight that, to eliminate the influence of machine component measurements, the displacement values for the experimental curves were adjusted using a correction factor of 0.6. This value was experimentally determined and represents an adjustment of the compliance of the testing machine and the used test setup. Figures 7 and 8 present the results for the vertical and inclined configurations, respectively.

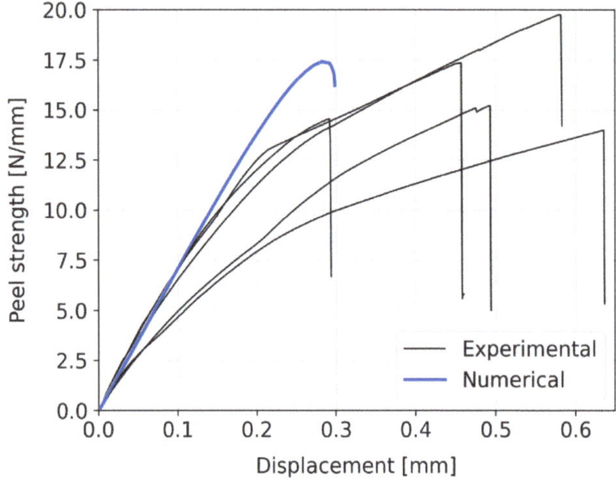

Figure 7. Comparison between experimental and numerical peel strength—displacement curves for the vertical configuration of the L-joint.

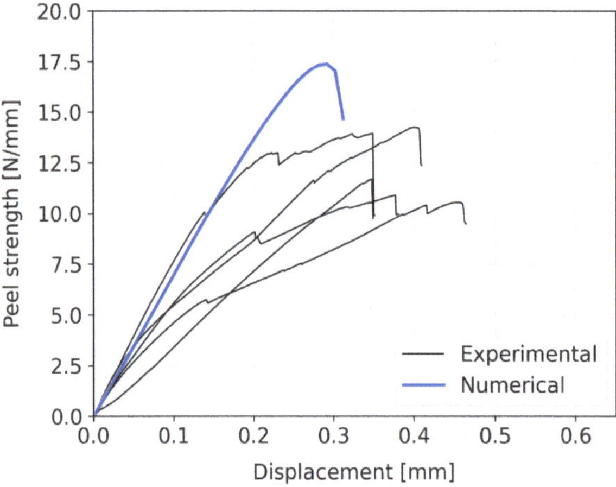

Figure 8. Comparison between experimental and numerical peel strength—displacement curves for the inclined configuration of the L-joint.

In the case of pine wood with a vertical fiber orientation of the upper substrate (Figure 7), the average peel strength of the joint was 16.2 ± 2.1 N/mm, with a corresponding displacement of 0.49 ± 0.12 mm. On the other hand, for the alternative geometry (Figure 8), the joint sustained an average peel strength of 12.2 ± 1.5 N/mm, resulting in a displacement of 0.38 ± 0.03 mm. The energy absorbed by the joints was determined by the area under the peel strength—displacement curves. The energy absorbed up to the point of failure for the vertical configuration was approximately 2.6 higher than the absorbed energy by the inclined configuration.

A comparison between these two cases reveals notable differences. Specifically, the first case exhibited approximately 25% higher peel strength than the second case, along with a superior displacement-to-failure capacity of around 22%. Furthermore, the observed behavior in the two cases significantly differs. In the vertical configuration (Figure 7), there is a noticeable shift in the material's behavior, marked by a distinct change in the slope of the curve. Initially, the material demonstrates elastic behavior, but it subsequently transitions to a plastic behavior, allowing for more displacement before failure occurs. On the other hand, in the inclined configuration (Figure 8), cracks in the wood are present, which were not observed in the vertical configuration. During the testing, the propagation of cracks could be heard, although they would not be visible. This indicated the presence and propagation of microcracks during testing, which would influence the fracture behavior of this particular configuration. These microcracks suggest the initiation of localized damage in the inclined configuration, contributing to its unique behavior. In conclusion, the two configurations not only vary in peel strength and displacement capacity but also exhibit distinct behaviors during testing.

Moreover, it is crucial to emphasize that the vertical configuration demonstrates superior characteristics including higher energy absorption, greater peel strength, and increased displacement capability prior to failure. The presence of vertical fiber orientations in the overlap region results in the loading being applied along the strongest direction, thereby enhancing the stiffness and peel strength of the joint. The vertical alignment of the fibers maximizes their load-bearing capacity, leading to improved overall performance of the joint.

The numerical computations conducted for both configurations are presented in form of resulting peel strength—displacement curves. These results were directly extracted and compared to the experimental data. Figure 7 illustrates the comparison between numerical and experimental results for the vertical configuration, while Figure 8 presents the corresponding comparison for the inclined configuration.

In the case of the vertical configuration, the numerical analysis simulated the peel strength of 17.4 N/mm with a displacement of 0.28 mm. In contrast, the experimental peel strength was reported as 16.2 ± 2.1 N/mm, accompanied by a displacement of 0.49 ± 0.12 mm. This indicates that the numerical model presents a reasonable agreement for the failure load. However, there is some deviation in displacement between the numerical and experimental results. Specifically, up to approximately 0.2 mm, the numerical and experimental results align well. Beyond this point, a noticeable divergence emerges in the behavior of the two sets of results. One possible explanation for this disparity is that the experimental data suggests signs of plastic behavior, allowing for greater displacement before failure. Unfortunately, the numerical simulation did not account for this phenomenon due to the lack of information to simulate such behavior. Despite the limitation of considering only the elastic behavior of wood, the numerical results still offer reasonably accurate predictions.

In the inclined configuration, the numerical analysis determined peel strength of 17.3 N/mm with a displacement of 0.29 mm. In contrast, experimental testing yielded a peel strength of 12.2 ± 1.5 N/mm, accompanied by a displacement of 0.38 ± 0.03 mm. As mentioned, in this configuration, the presence of microcracks in the experimental testing was noted, contributing to the observed discrepancy, since they are not considered in the finite element analysis (FEA). Consequently, the FEA tends to predict higher failure loads compared to the experimental results, proving to have a worse performance for this configuration in terms of the failure load. However, a smaller discrepancy in the final displacement between numerical and experimental data was observed when compared to the previous configuration.

Comparing both numerical results shows strikingly similar patterns in terms of the failure load, displacement, and the peel strength—displacement curve shape. However, real-world behavior diverges. The experimental data are scattered, and the authors believe that the source of these dispersion is mainly caused by the wood itself and thus a key limitation associated with using these materials. Even when sourced from the same specimen, the different locations along the wood can have drastically varied behavior. In a design-based approach, one must find ways to avoid having to precisely characterize these geometrical and material uncertainties and try to employ a more general approach. In the vertical setup, the numerical and experimental results closely match on the failure load, though not on the displacement. The numerical model does not consider plasticity, unlike the experiments, which show more displacement tolerance. In the inclined case, discrepancies arise due to microcracks unaccounted for in the numerical analysis. Despite these differences, the numerical model remains a valuable tool for predicting joint strength, particularly within the elastic range.

3.2. Failure Mode

Following the testing phase, a detailed analysis was conducted to identify the type of failure modes observed for both cases. Two types of behavior were registered. In both cases, a visibly non-reflective surface in the overlap region strongly suggests the occurrence of delamination. In the case of the vertical configuration, the delamination process involved a greater number of fibers being pulled, accompanied by a fracture occurring within the wood at a very oblique angle (Figure 9a). Conversely, for the inclined configuration, the plane of failure exhibited a more vertical orientation and was sudden, resulting in a flatter failure surface (Figure 9b). All of the tested specimens failed as presented in Figure 9.

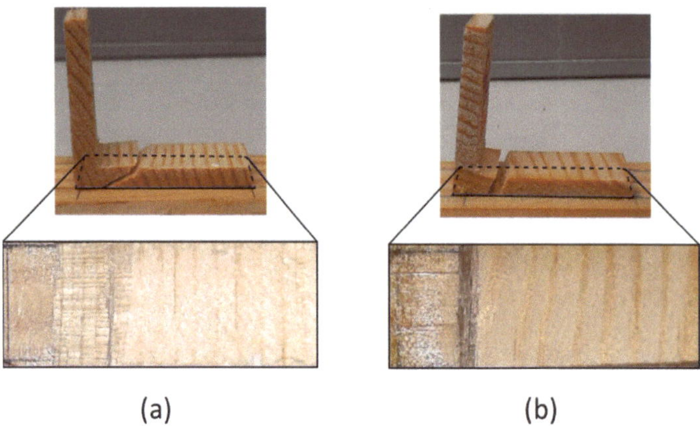

(a)   (b)

Figure 9. Failure mode observation for (**a**) vertical configuration and (**b**) inclined configuration.

In the numerical analysis, delamination was predicted as the predominant failure mode, and it was imperative to understand this phenomenon. The monitoring of damage progression utilized the state variable SDEG (Figure 10), which corresponds to stiffness degradation of the cohesive element, with zero indicating the absence of damage, and a linear increase in SDEG signifying damage progression until reaching one, resulting in the failure of the cohesive element.

To address the discrepancies observed between the numerical model and experimental observations, and considering a global approach to the problem, a thorough analysis was conducted, leading to the consideration of the maximum principal stress criteria, specifically the maximum principal stress failure predictor (MPSFP). According to MPSFP, a fracture occurs when the local strength is exceeded by the maximum principal stress in a multiaxial stress system. Using this approach, it was possible to check, for a given displacement, the plane in which the maximum principal stress reached a value greater than the material's strength. As shown in Figure 11, although in general the specimen is below the strength of the RL plane of 16 MPa, the red arrow indicates that locally, in this plane, this value is exceeded, with 17 MPa and the orientation represented by the same arrow. Since a crack propagates in the direction perpendicular to the plane with the maximum principal stress, this numerical simulation allows us to obtain the correct orientation of the crack propagation when compared to the experimental results. Comparing the images in Figure 9 with the results in Figure 11, a good agreement can be observed.

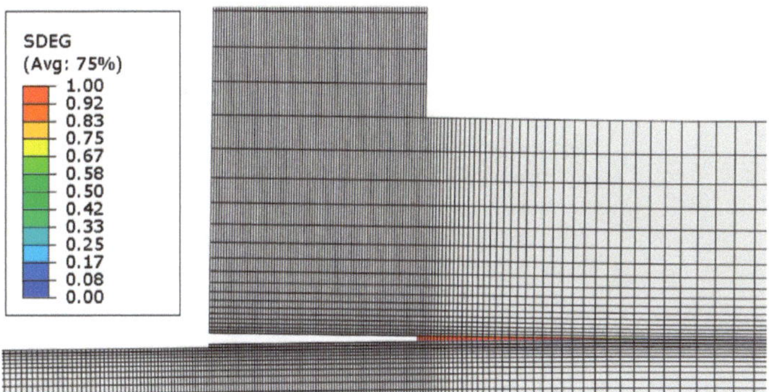

Figure 10. Damage analysis (SDEG) for L-joint at failure.

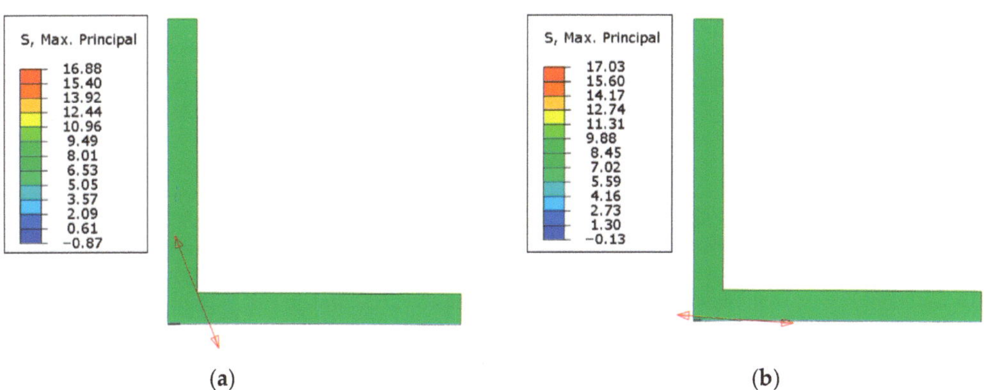

(a)
(b)

Figure 11. Maximum principal stress distribution in the FEA: (**a**) vertical configuration and (**b**) inclined configuration, where the red arrows indicate the local direction where the maximum principal stress was attained.

4. Conclusions

This study focused on investigating L-joints for the automotive industry by incorporating biomaterials, specifically pine wooden substrates with a novel polyurethane-based bioadhesive. Two different configurations using pine wood were analyzed through strength analysis and failure mode. This study involved experimental and numerical analyses, which were compared to each other.

In comparing the two configurations, significant distinctions emerged. The first configuration demonstrated a notable 25% increase in strength and a 22% improvement in displacement-to-failure capacity compared to the second. These differences were accompanied by varying material behaviors during testing. In the vertical setup, the material initially exhibited elastic behavior but shifted to plastic behavior, allowing for more displacement before failure. Conversely, the inclined configuration indicated the presence of microcracks, which suggests a localized damage not observed in the vertical setup. These findings highlight differences not only in strength and displacement capacity but also in observed behaviors. The vertical configuration displayed superior characteristics, including higher energy absorption, increased strength, and enhanced displacement capability before failure. This superiority stemmed from vertical fiber orientations in the overlap region, facilitating loading along the material's strongest direction. The analysis of failure modes identified delamination as the predominant issue in both cases, with visible non-reflective surfaces indicating this phenomenon. In the vertical configuration, delamination involved more fibers being pulled, resulting in an oblique fracture within the wood. In contrast, the inclined configuration exhibited a more vertically oriented failure, leading to a flatter failure surface.

Regarding numerical versus experimental results, the numerical analysis presented a reasonable agreement with experimental data in terms of the failure load for the vertical configuration. However, deviations in displacement were observed, primarily due to the unaccounted plastic behavior in the numerical simulations. Regarding the inclined configuration, during the testing of the inclined configuration, the propagation of cracks could be heard, although they were not visible. Thus, the authors consider that the propagation of microcracks during testing was an influencing parameter in the fracture behavior of this particular configuration, contributing to the different behavior observed in the numerical model.

To address these discrepancies, this study introduced the maximum principal stress criteria, specifically the maximum principal stress failure predictor (MPSFP). MPSFP stipulates that fracture occurs when the maximum principal stress surpasses local strength in a multiaxial stress system. The verification revealed that, at a specific displacement, the maximum principal stress exceeded local strength for the RL plane, explaining the observed crack propagation. Although there was an attempt to use a simple model, which is able to reproduce the behavior of the wood without the need for highly complex characterization procedures, in future works, the present study should be expanded to create a more accurate model for the bonded wood. Furthermore, it will be of great interest to use these same models to study and predict the behavior of these structures considering the effects of aging and dynamic loads. In fact, one of the main problems facing these wooden structures is that their long-term behavior is not yet fully understood or modeled.

In summary, this study underscores differences in behavior between configurations, acknowledging the reliability of numerical modeling while recognizing limitations related to plasticity and microcracks. It also emphasizes the significance of considering failure modes and criteria for accurate simulations.

Author Contributions: Conceptualization, A.M.S.C.; Methodology, A.M.S.C., C.S.P.B., S.J., E.A.S.M., R.J.C.C., T.V. and L.F.M.d.S.; Software, A.M.S.C. and B.D.S.; Validation, A.M.S.C., C.S.P.B., E.A.S.M., R.J.C.C., T.V. and L.F.M.d.S.; Formal analysis, A.M.S.C., C.S.P.B. and S.J.; Investigation, A.M.S.C. and S.J.; Resources, J.C.B. and L.F.M.d.S.; Data curation, A.M.S.C. and B.D.S.; Writing—original draft, A.M.S.C.; Writing—review & editing, C.S.P.B., S.J., B.D.S., E.A.S.M., R.J.C.C., T.V. and L.F.M.d.S.; Supervision, C.S.P.B., B.D.S., E.A.S.M., R.J.C.C., J.C.B. and L.F.M.d.S.; Project administration, L.F.M.d.S.; Funding acquisition, L.F.M.d.S. All authors have read and agreed to the published version of the manuscript.

Funding: This research was funded by the Project No. PTDC/EME-EME/6442/2020 "A smart and eco-friendly adhesively bonded structure for the next generation mobility platforms" and the individual grants 2022.12426.BD and CEECIND/03276/2018, funded by national funds through the Portuguese Foundation for Science and Technology (FCT).

Institutional Review Board Statement: Not applicable.

Data Availability Statement: The raw/processed data required to reproduce these findings cannot be shared at this time as the data are part of an ongoing study.

Conflicts of Interest: The authors declare no conflict of interest.

References

1. Da Silva, L.F.M.; Öchsner, A.; Adams, R.D. *Handbook of Adhesion Technology*; Springer: Berlin/Heidelberg, Germany, 2018; Volume 1.
2. Borges, C.S.P.; Jalali, S.; Tsokanas, P.; Marques, E.A.S.; Carbas, R.J.C.; da Silva, L.F.M. Sustainable Development Approaches through Wooden Adhesive Joints Design. *Polymers* **2022**, *15*, 89. [CrossRef] [PubMed]
3. Bachmann, J.; Yi, X.; Gong, H.; Martinez, X.; Bugeda, G.; Oller, S.; Tserpes, K.; Ramon, E.; Paris, C.; Moreira, P.; et al. Outlook on ecologically improved composites for aviation interior and secondary structures. *CEAS Aeronaut. J.* **2018**, *9*, 533–543. [CrossRef]
4. Lanvermann, C.; Evans, R.; Schmitt, U.; Hering, S.; Niemz, P. Distribution of structure and lignin within growth rings of Norway spruce. *Wood Sci. Technol.* **2013**, *47*, 627–641. [CrossRef]
5. Xu, Y.; Han, Y.; Li, Y.; Li, J.; Li, J.; Gao, Q. Preparation of a strong, mildew-resistant, and flame-retardant biomimetic multi-functional soy protein adhesive via the construction of an organic-inorganic hybrid multiple-bonding structure. *Chem. Eng. J.* **2022**, *437*, 135437. [CrossRef]
6. Jalali, S.; Borges, C.D.S.P.; Carbas, R.J.C.; Marques, E.A.D.S.; Akhavan-Safar, A.; Barbosa, A.S.O.F.; Bordado, J.C.M.; da Silva, L.F.M. A Novel Technique for Substrate Toughening in Wood Single Lap Joints Using a Zero-Thickness Bio-Adhesive. *Materials* **2024**, *17*, 448. [CrossRef] [PubMed]
7. Islam, N.; Rahman, F.; Das, A.K.; Hiziroglu, S. An overview of different types and potential of bio-based adhesives used for wood products. *Int. J. Adhes. Adhes.* **2021**, *112*, 102992. [CrossRef]
8. Yuan, C.; Chen, M.; Luo, J.; Li, X.; Gao, Q.; Li, J. A novel water-based process produces eco-friendly bio-adhesive made from green cross-linked soybean soluble polysaccharide and soy protein. *Carbohydr. Polym.* **2017**, *169*, 417–425. [CrossRef]
9. Keplinger, T.; Wittel, F.K.; Rüggeberg, M.; Burgert, I. Wood derived cellulose scaffolds—Processing and mechanics. *Adv. Mater.* **2021**, *33*, 2001375. [CrossRef]
10. Ferdosian, F.; Pan, Z.; Gao, G.; Zhao, B. Bio-based adhesives and evaluation for wood composites application. *Polymers* **2017**, *9*, 70. [CrossRef]
11. Shang, X.; Marques, E.; Machado, J.; Carbas, R.; Jiang, D.; da Silva, L. Review on techniques to improve the strength of adhesive joints with composite adherends. *Compos. Part B Eng.* **2019**, *177*, 107363. [CrossRef]
12. Tserpes, K.; Barroso-Caro, A.; Carraro, P.A.; Beber, V.C.; Floros, I.; Gamon, W.; Kozłowski, M.; Santandrea, F.; Shahverdi, M.; Skejić, D.; et al. A review on failure theories and simulation models for adhesive joints. *J. Adhes.* **2022**, *98*, 1855–1915. [CrossRef]
13. Santoni, I.; Pizzo, B. Evaluation of alternative vegetable proteins as wood adhesives. *Ind. Crop. Prod.* **2013**, *45*, 148–154. [CrossRef]
14. Borges, C.; Nunes, P.; Akhavan-Safar, A.; Marques, E.; Carbas, R.; Alfonso, L.; Silva, L. A strain rate dependent cohesive zone element for mode I modeling of the fracture behavior of adhesives. *Proc. Inst. Mech. Eng. Part L J. Mater. Des. Appl.* **2020**, *234*, 610–621. [CrossRef]
15. Oliveira, P.R.; May, M.; Panzera, T.H.; Scarpa, F.; Hiermaier, S. Reinforced biobased adhesive for eco-friendly sandwich panels. *Int. J. Adhes. Adhes.* **2020**, *98*, 102550. [CrossRef]
16. de Moura, M.; Campilho, R.; Gonçalves, J. Pure mode II fracture characterization of composite bonded joints. *Int. J. Solids Struct.* **2009**, *46*, 1589–1595. [CrossRef]
17. De Moura, M.F.S.F.; Dourado, N. *Wood Fracture Characterisation*; CRC Press: Boca Raton, FL, USA, 2018.
18. Pečnik, J.G.; Gavrić, I.; Sebera, V.; Kržan, M.; Kwiecień, A.; Zając, B.; Azinović, B. Mechanical performance of timber connections made of thick flexible polyurethane adhesives. *Eng. Struct.* **2021**, *247*, 113125. [CrossRef]
19. Rocha, A.; Akhavan-Safar, A.; Carbas, R.; Marques, E.; Goyal, R.; El-Zein, M.; da Silva, L. Numerical analysis of mixed-mode fatigue crack growth of adhesive joints using CZM. *Theor. Appl. Fract. Mech.* **2020**, *106*, 102493. [CrossRef]
20. Tabiei, A.; Wu, J. Three-dimensional nonlinear orthotropic finite element material model for wood. *Compos. Struct.* **2000**, *50*, 143–149. [CrossRef]
21. Moses, D.; Prion, H. Stress and failure analysis of wood composites: A new model. *Compos. Part B Eng.* **2004**, *35*, 251–261. [CrossRef]
22. Díaz, J.d.C.; Nieto, P.G.; Martínez-Luengas, A.L.; Domínguez, F.S.; Hernández, J.D. Non-linear numerical analysis of plywood board timber connections by DOE-FEM and full-scale experimental validation. *Eng. Struct.* **2013**, *49*, 76–90. [CrossRef]
23. Nunes, P.D.; Borges, C.S.; Marques, E.A.; Carbas, R.J.; Akhavan-Safar, A.; Antunes, D.P.; Lopes, A.M.; da Silva, L.F. Numerical assessment of strain rate in an adhesive layer throughout double cantilever beam and end notch flexure tests. *Proc. Inst. Mech. Eng. Part E J. Process. Mech. Eng.* **2020**, *234*, 415–425. [CrossRef]
24. Hofstetter, K.; Gamstedt, E.K. Hierarchical Modelling of Microstructural Effects on Mechanical Properties of Wood. A Review COST Action E35 2004–2008: Wood Machining–Micromechanics and Fracture. 2009. Available online: https://www.degruyter.com/document/doi/10.1515/HF.2009.018/html (accessed on 1 July 2023).
25. Kandler, G.; Füssl, J.; Eberhardsteiner, J. Stochastic finite element approaches for wood-based products: Theoretical framework and review of methods. *Wood Sci. Technol.* **2015**, *49*, 1055–1097. [CrossRef]

26. Kandler, G.; Füssl, J.; Serrano, E.; Eberhardsteiner, J. Effective stiffness prediction of GLT beams based on stiffness distributions of individual lamellas. *Wood Sci. Technol.* **2015**, *49*, 1101–1121. [CrossRef]
27. Mackenzie-Helnwein, P.; Müllner, H.W.; Eberhardsteiner, J.; Mang, H.A. Analysis of layered wooden shells using an orthotropic elasto-plastic model for multi-axial loading of clear spruce wood. *Comput. Methods Appl. Mech. Eng.* **2005**, *194*, 2661–2685. [CrossRef]
28. Méité, M.; Dubois, F.; Pop, O.; Absi, J. Mixed mode fracture properties characterization for wood by digital images correlation and finite element method coupling. *Eng. Fract. Mech.* **2013**, *105*, 86–100. [CrossRef]
29. Berg, S.; Sandberg, D.; Ekevad, M.; Vaziri, M. Crack influence on load-bearing capacity of glued laminated timber using extended finite element modelling. *Wood Mater. Sci. Eng.* **2015**, *10*, 335–343. [CrossRef]
30. Huang, W.; Sun, L.; Li, L.; Shen, L.; Huang, B.; Zhang, Y. Investigations on low-energy impact and post-impact fatigue of adhesively bonded single-lap joints using composites substrates. *J. Adhes.* **2019**, *96*, 1326–1354. [CrossRef]
31. López-Puente, J.; Arias, A.; Zaera, R.; Navarro, C. The effect of the thickness of the adhesive layer on the ballistic limit of ceramic/metal armours. An experimental and numerical study. *Int. J. Impact Eng.* **2005**, *32*, 321–336. [CrossRef]
32. Corte-Real, L.M.R.M.; Jalali, S.; Borges, C.S.P.; Marques, E.A.S.; Carbas, R.J.C.; da Silva, L.F.M. Development and Characterisation of Joints with Novel Densified and Wood/Cork Composite Substrates. *Materials* **2022**, *15*, 7163. [CrossRef]
33. Jalali, S.; Borges, C.D.S.P.; Carbas, R.J.C.; Marques, E.A.D.S.; Bordado, J.C.M.; da Silva, L.F.M. Characterization of Densified Pine Wood and a Zero-Thickness Bio-Based Adhesive for Eco-Friendly Structural Applications. *Materials* **2023**, *16*, 7147. [CrossRef]

polymers

Article

A Study of Hydroxyl-Terminated Block Copolyether-Based Binder Curing Kinetics

Wu Yang [1], Zhengmao Ding [2], Cong Zhu [1], Tianqi Li [1], Wenhao Liu [1] and Yunjun Luo [1,*]

[1] School of Materials Science and Engineering, Beijing Institute of Technology, Beijing 100081, China
[2] Pen-Tung Sah Institute of Micro-Nano Science and Technology, Xiamen University, Xiamen 361005, China
* Correspondence: yjluo@bit.edu.cn

Abstract: In order to determine the curing reaction model and corresponding parameters of hydroxyl-terminated block copolyether (HTPE) and provide a theoretical reference for its practical application, the non-isothermal differential scanning calorimetry (DSC) method was used to analyze the curing processes of three curing systems with HTPE and N-100 (an aliphatic polyisocyanate curing agent), isophorone diisocyanate (IPDI), and a mixture of N-100 and IPDI as curing agents. The results show that the curing activation energy of N-100 and HTPE was about 69.37 kJ/mol, slightly lower than the curing activation energy of IPDI and HTPE (75.60 kJ/mol), and the curing activation energy of the mixed curing agent and HTPE was 69.79 kJ/mol. The curing process of HTPE conformed to the autocatalytic reaction model. The non-catalytic reaction order (n) of N-100 and HTPE was about 1.2, and the autocatalytic order (m) was about 0.3, both lower than those of IPDI and HTPE. The reaction kinetics parameters of the N-100 and IPDI mixed curing agent with HTPE were close to those of N-100 and HTPE. The verification results indicate a high degree of overlap between the experimental data and the calculated data.

Keywords: HTPE; curing kinetics; non-isothermal DSC; isoconversional methods; autocatalytic reaction model

check for
updates

Citation: Yang, W.; Ding, Z.; Zhu, C.; Li, T.; Liu, W.; Luo, Y. A Study of Hydroxyl-Terminated Block Copolyether-Based Binder Curing Kinetics. *Polymers* 2024, 16, 2246. https://doi.org/10.3390/polym16162246

Academic Editor: Nicolas Sbirrazzuoli

Received: 7 July 2024
Revised: 4 August 2024
Accepted: 5 August 2024
Published: 7 August 2024

1. Introduction

Hydroxyl-terminated block copolyether (HTPE), as a block copolymer ether composed of polyethylene glycol (PEG) and polytetrafluoroethylene (PTMG), is considered to be a new type of binder that is expected to replace hydroxyl-terminated polybutadiene (HTPB) in the field of composite solid propellants [1] due to its excellent performance, particularly its outstanding low vulnerability characteristics [2]. At present, a large number of works on the mechanics [3,4], thermal decomposition [5,6], combustion [7–9], and safety performance [10,11] of HTPE-based propellants have been published, and numerous studies have shown that HTPE has good application prospects in the field of composite solid propellants.

However, there are few reports on the curing kinetics of HTPE. The curing reaction between HTPE and the curing agent is directly related to the process parameters, such as the pot life and curing time of the propellant, and is an important step in the preparation process of composite solid propellants [12]. Therefore, studying the curing kinetics of HTPE helps to better understand its curing reaction process and is very helpful for the specific application of HTPE [13].

Many tests can be used to analyze the curing kinetics, such as low-field nuclear magnetic resonance testing [14], rheological testing [15,16], infrared testing [17,18], differential scanning calorimetry (DSC) testing [19–21], etc. Among them, the DSC method has the characteristics of high sensitivity, simple operation, and a short testing cycle and has been widely used. DSC testing can detect and record the heat changes during the curing reaction process. By mathematically processing and analyzing the recorded data, researchers can obtain the reaction model and kinetic parameters of the curing reaction [22]. It should be noted that there are two commonly used methods for the study of curing kinetics parameters using DSC, namely the isothermal method and the non-isothermal method. These two methods each have their own characteristics and scope. Among them, the isothermal method tests the curing reaction of a sample at multiple fixed temperatures (usually at least four different temperatures), records the change in heat release during the curing reaction process with the reaction time until the end of the reaction, and analyzes it. This method can test the curing reaction of the sample at ambient temperature during its actual application, which is more in line with the actual application situation of the sample. However, its disadvantage is that when the curing reaction of the sample is slow and the heat released during the reaction is low, it can lead to a long testing cycle and the inaccurate detection of the instrument in the later stage of the curing reaction due to the low heat generated by the reaction, resulting in serious data errors. This is unacceptable for the analysis of curing reaction kinetics. Therefore, the isothermal DSC method is commonly used to study reactions with high heat release and fast reaction rates, such as photopolymerization reactions. However, the curing reaction of HTPE, studied in this work, does not meet the requirements. On the other hand, the non-isothermal DSC method records the variation in the heat release of the sample during the curing reaction with the temperature at different heating rates (usually at least four different heating rates) within a certain temperature range, until the curing reaction is complete, and analyzes it. Compared to the isothermal DSC method, the non-isothermal DSC method has a shorter testing period and more accurate data, making it more commonly used in reaction kinetics analysis. This is also the method applied in this study. It is worth noting that whether using the isothermal DSC method or the non-isothermal DSC method, it should be ensured that only the curing reaction of the sample produces thermal effects during the testing process, and there are no other processes (such as volatilization, thermal decomposition, etc.), because the use of the DSC method to study the curing reaction kinetics is based on the reaction heat data recorded by DSC. Otherwise, it will result in inaccuracies or the impossibility of analyzing the curing kinetics. Therefore, it is necessary to choose an appropriate temperature or heating rate.

Moreover, the principle of the HTPE curing reaction is that the hydroxyl groups at the ends of the molecular chains of the HTPE prepolymer react with the isocyanate groups in the curing agent to form amino ester groups. During the reaction, the molecular chains expand to form a three-dimensional crosslinked network, transforming the reactants from a liquid to a solid state. Therefore, the curing reaction of HTPE requires the participation of curing agents with isocyanate groups and a minimum functionality of 2; otherwise, the expansion of the molecular chains cannot be achieved. There are many types of curing agents that meet these requirements, of which N-100 (an aliphatic polyisocyanate curing agent) is a commonly used curing agent for the curing reaction of HTPE due to its functionality being greater than 3, making it easy to form chemical crosslinking points [23–25]. On the other hand, isophorone diisocyanate (IPDI) is a commonly used curing agent for polyurethane curing reactions [26,27], and it has also been applied in the HTPE system [28]. Therefore, in this study, to determine the reaction model and corresponding kinetic parameters of the HTPE curing reaction, and to provide certain references for the practical application of HTPE, the non-isothermal DSC method was used to analyze the curing kinetics of HTPE with N-100, IPDI, and mixtures of N-100 and IPDI. Finally, it was found that the curing reaction process of HTPE with N-100 and IPDI was consistent with the autocatalytic reaction model, and the corresponding kinetic parameters were obtained by calculation and fitting.

2. Materials and Methods

2.1. Materials

Hydroxyl-terminated block copolyether (HTPE) and N-100 (an aliphatic polyisocyanate curing agent) were supplied by the Liming Research Institute of Chemical Industry,

Luoyang, China. Isophorone diisocyanate (IPDI) was obtained from Aladdin. Dioctyl sebacate (DOS) was analytically pure and obtained from the Tianjin Guangfu Fine Chemical Research Institute. Triphenyl bismuth (TPB) (purity of 99%) and dibutyltin dilaurate (DBTDL) were obtained from the Shanghai Institute of Organic Chemistry (Shanghai Municipality, China) and were formulated into a 0.5 wt.% solution with DOS as the solvent and mixed in a mass ratio of TPB/DBTDL of 3:1 before being used as the curing catalyst [29].

2.2. Non-Isothermal Differential Scanning Calorimetry Tests

Non-isothermal differential scanning calorimetric (DSC) analysis was carried out in the DSC 3 (METTLER TOLEDO). Approximately 5~8 mg of the reaction mixture, comprising HTPE and the curing agent at a stoichiometric ratio of 1, was placed in an aluminum crucible. Before placing the mixture into the crucible, it needed to be thoroughly stirred to ensure that the curing agent, catalyst, and binder were mixed evenly. Then, the mixture was subjected to a heating schedule up to 473 K from 298 K at various heating rates (1, 2, 3, and 4 K/min) under an atmosphere of nitrogen, purged at a rate of 40 mL/min. The curing agents used were N-100, IPDI, and a mixed curing agent composed of N-100 and IPDI. According to previous experiments, when the mass ratio of N-100 to IPDI was 4:1, the prepared adhesive had good performance. Therefore, in the curing kinetics study in this work, the mass ratio of N-100 to IPDI contained in the mixed curing agent used was 4:1. The curing system was named N-100/HTPE, IPDI/HTPE, or NI/HTPE according to the curing agent used.

3. Results

3.1. Curing Activation Energy

The heat flow curves of N-100/HTPE, IPDI/HTPE, and NI/HTPE are shown in Figure 1, and the corresponding conversion rate (α) versus temperature curves are also displayed. It is worth noting that different testing methods correspond to different calculation methods for the conversion rates. For non-isothermal DSC methods, calculating the conversion rates requires the integration of the heat flow curves at different heating rates to obtain the total heat release of the curing reaction at each heating rate. Then, the accumulated heat release at a certain time during the reaction process can be compared with the total heat release of the entire reaction to obtain the conversion rate α at that time. At the beginning of the reaction, α is 0, while, at the end of the reaction, α is 1. This is because, when analyzing the reaction kinetics using the DSC method, it is believed that only the reactants will produce thermal effects when they undergo a reaction—that is, the beginning and end of the exothermic reaction are used as indicators of the start and end of the reaction, and the exothermic process represents the reaction process.

From Figure 1, it can be seen that the heat flow curves of the three curing systems all have only one exothermic peak. As the heating rate increases, the peak temperature ($T_{\max}$) also increases, and the shape of the exothermic peak remains almost unchanged but becomes higher and sharper. This is a common phenomenon in non-isothermal DSC testing analysis. The reason is that, in the non-isothermal DSC testing process, the test sample is heated by the instrument, and the heating rate is controlled by the program. However, in reality, the temperature distribution inside the sample is uneven. The faster the heating rate, the more uneven the temperature distribution inside the sample and the greater the temperature difference between the inside and outside, which leads to an increase in the peak temperature. At the same time, an increase in the heating rate will also cause the sample to reach a higher temperature faster, resulting in an increase in the peak temperature with an increase in the heating rate. Moreover, an increase in the heating rate will lead to an increase in the thermal effect per unit of time, resulting in greater temperature changes and an increase in the peak height. Meanwhile, the conversion rates of the three curing systems exhibit a typical S-shaped curve as a function of the temperature, and the slope of the curve increases with the increase in the heating rate.

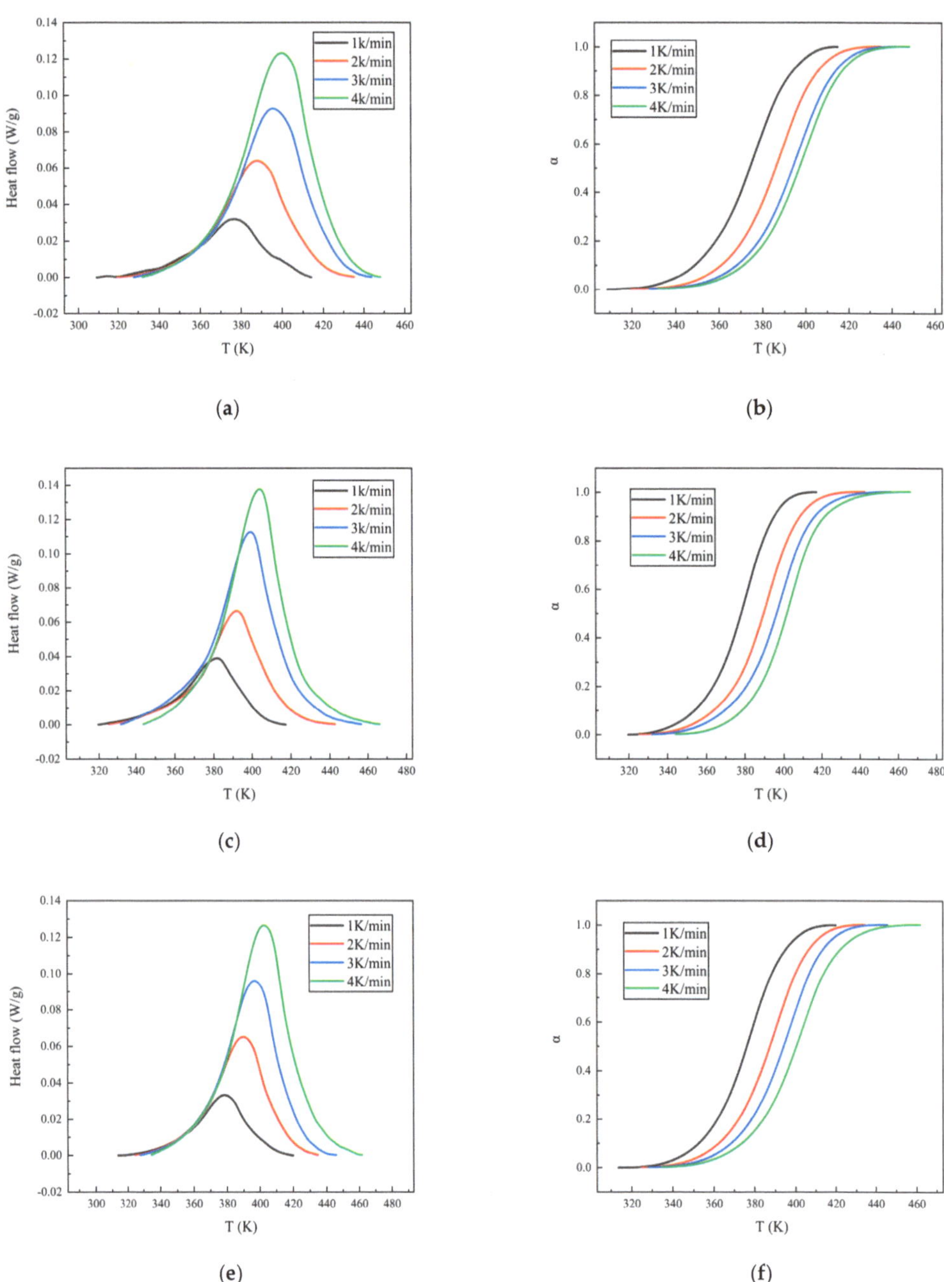

Figure 1. (**a**) The heat flow curves of N-100/HTPE, (**b**) the α versus temperature curves of N-100/HTPE, (**c**) the heat flow curves of IPDI/HTPE, (**d**) the α versus temperature curves of IPDI/HTPE, (**e**) the heat flow curves of NI/HTPE, and (**f**) the α versus temperature curves of NI/HTPE. The colors of the curves in each graph represent the heating rates, with black, red, blue, and green representing heating rates of 1 K/min, 2 K/min, 3 K/min, and 4K/min, respectively.

Equations (1) and (2) represent the Kissinger model [30] and the Ozawa model [31], respectively. It can be seen from them that the curing activation energy (E_a) can be calculated based on the T_{max} of the heat flow curves at different heating rates. The T_{max} and activation energies of the three curing systems at different heating rates are listed in Table 1. Figure 2 shows the fitting results of the three curing systems with the Kissinger and Ozawa models.

$$\frac{d\ln\left(\beta / T_{max}^2\right)}{d(1/T_{max})} = -\frac{E_a}{R} \tag{1}$$

$$\frac{d\ln \beta}{d(1/T_{max})} = -1.052\frac{E_a}{R} \tag{2}$$

where β is the heating rate, in K/min; R is the gas constant, with a value of 8.314 J/(mol·K).

Table 1. T_{max} and E_a of the three curing systems.

Curing System	Peak Temperature T_{max} (K)				E_a (kJ/mol)	
	1 K/min	2 K/min	3 K/min	4 K/min	Kissinger	Ozawa
N-100/HTPE	376.753	388.952	395.251	400.745	66.53	69.37
IPDI/HTPE	381.872	392.055	399.711	403.886	73.01	75.60
NI/HTPE	378.736	390.987	396.699	403.084	66.94	69.79

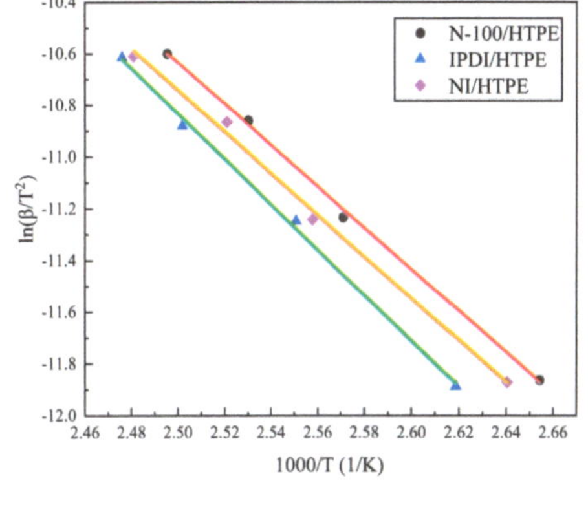
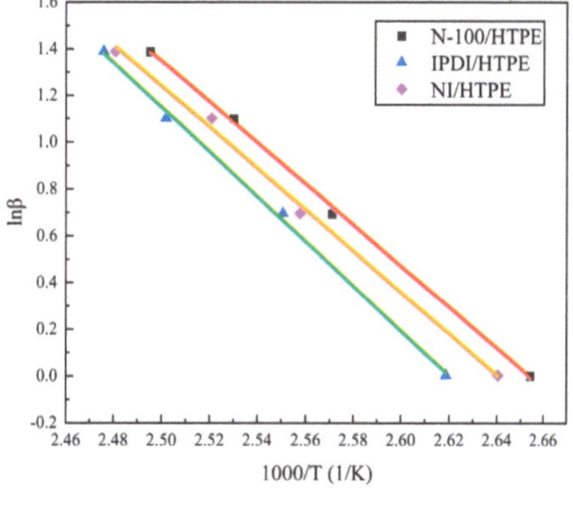

(a)

(b)

Figure 2. (**a**) The fitting results of the activation energy for the three curing systems using the Kissinger method and (**b**) the fitting results of the activation energy for the three curing systems using the Ozawa method. The dots in the figure represent the experimental data, the lines represent the fitting results, and the colors of the lines represent the different curing systems, where red represents N-100/HTPE, green represents IPDI/HTPE, and brown represents NI/HTPE.

Table 1 shows that the order of the peak temperatures for the three curing systems at various heating rates is N-100/HTPE < NI/HTPE < IPDI/HTPE, and the order of the activation energies is also the same. This indicates that, compared to IPDI, N-100 is more likely to react with HTPE, but the difference in reaction activity between the two curing agents is very minimal. From Figure 1, it can also be seen that, like N-100/HTPE and IPDI/HTPE, NI/HTPE only has one exothermic peak, and it cannot be divided into two or more through fitting. Therefore, it is reasonable to believe that both the reaction between N-100 and HTPE and the reaction between IPDI and HTPE occur simultaneously during the curing of the NI/HTPE system. It should be noted that the activation energies calculated using the Kissinger method and Ozara method are fixed values, which are not accurate enough and only serve as a reference, since the activation energy of the curing reaction is constantly changing with the progress of the reaction in reality.

The isoconversional method is a commonly used method for the calculation of the change in activation energy during the reaction process. Among them, isoconversional methods such as the Kissinger–Akahira–Sunose (KAS) method [32] and Flynn–Wall–Ozawa (FWO) method [33], as extensions of the Kissinger method and Ozawa method, respectively, are also commonly used to study the changes in activation energy during the reaction process. According to the conversion rate–temperature curves of the three curing systems at different heating rates, as displayed in Figure 1, the activation energy during the reaction process can be calculated using the isoconversional method. Figures 3–5 show the calculation results of the activation energies of N-100/HTPE, IPDI/HTPE, and NI/HTPE using the KAS method and FWO method, respectively.

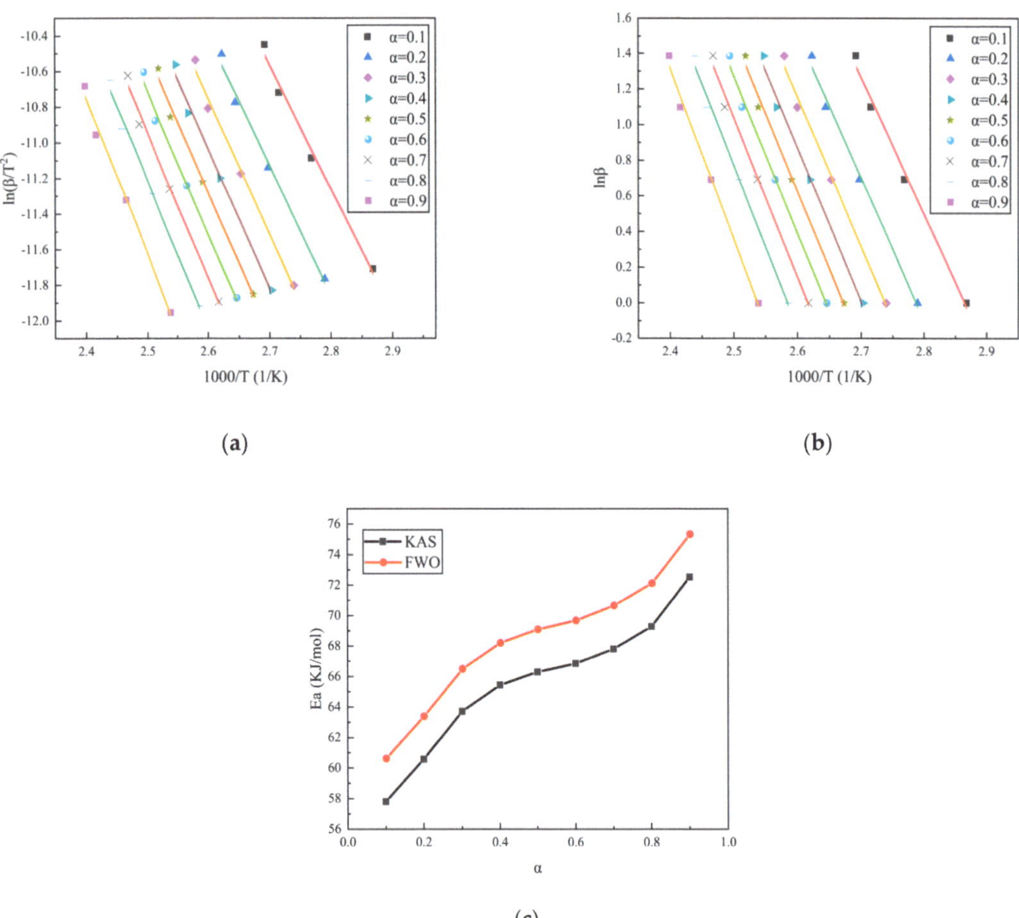

Figure 3. (**a**) The fitting results of N-100/HTPE based on the KAS method, (**b**) the fitting results of N-100/HTPE based on the FWO method, and (**c**) the activation energy calculation results of N-100/HTPE based on the KAS and FWO methods.

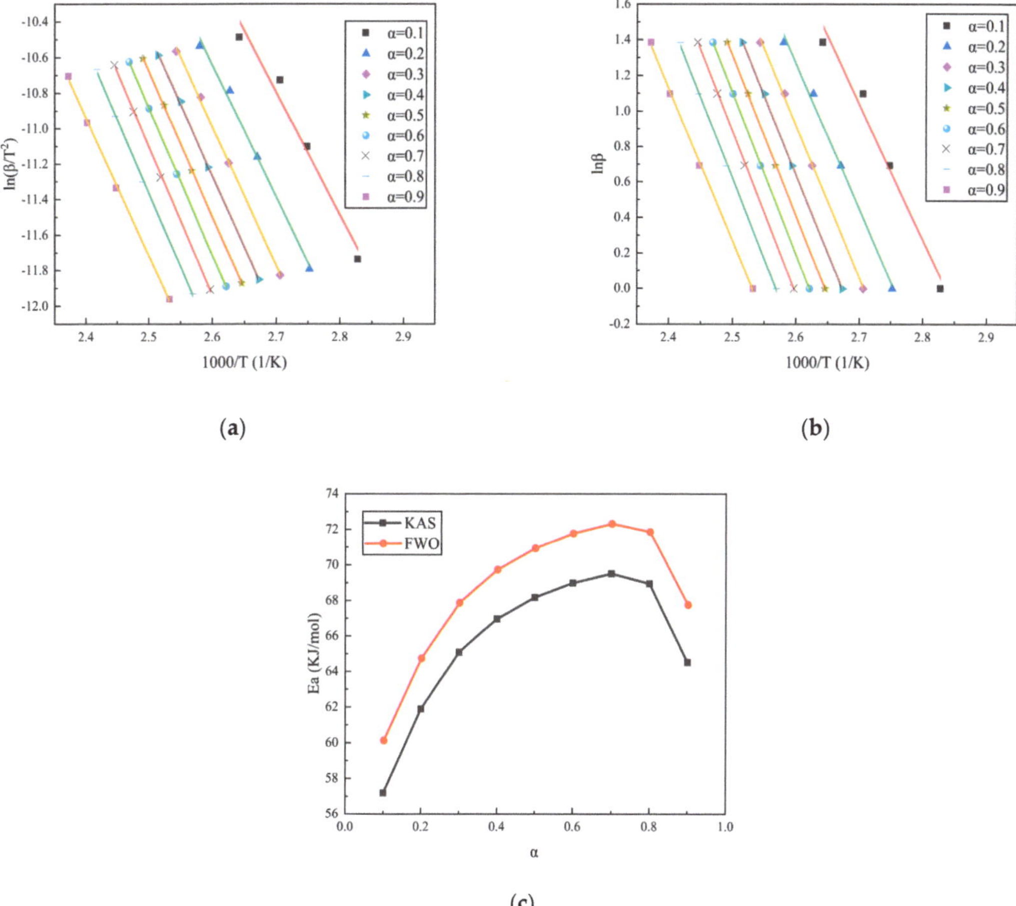

Figure 4. (**a**) The fitting results of IPDI/HTPE based on the KAS method, (**b**) the fitting results of IPDI/HTPE based on the FWO method, and (**c**) the activation energy calculation results of IPDI/HTPE based on the KAS and FWO methods.

From Figures 3–5, it can be seen that the changes in the activation energies calculated by the KAS and FWO methods have the same trends, and although the activation energies calculated by the KAS and FWO methods are different, the difference is very small. Similarly, the activation energies calculated by the Kissinger and Ozawa methods in Table 1 are also minimal. This is understandable as the KAS and FWO methods are extensions of the Kissinger and Ozawa methods, respectively. Furthermore, it can be observed that the activation energy changes of the IPDI/HTPE system are similar to those in reference [34], indicating that there is a significant diffusion control process in the later stage of the curing reaction. Through comparison, it can be found that the NI/HTPE system also exhibits similar phenomena, which should be attributed to the role of IPDI in the curing agent. On the other hand, as the reaction progresses, the concentrations of the reactants decrease, resulting in a decrease in the effective collision frequency between the reactants. This is also one of the reasons that the activation energy increases as the reaction progresses.

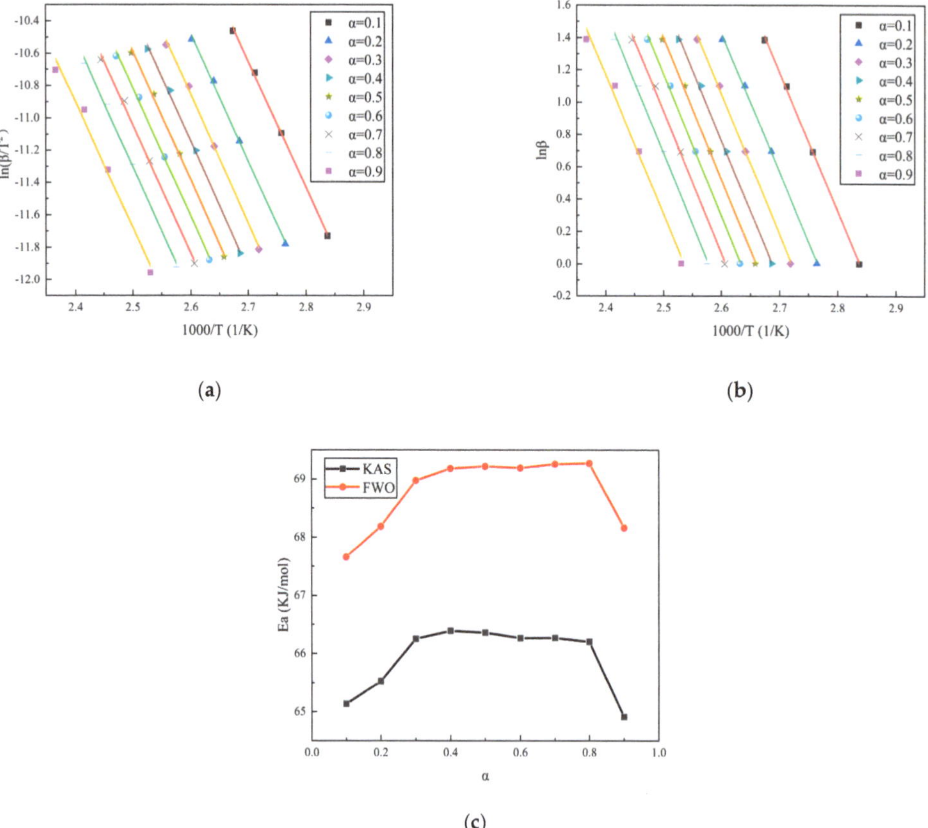

Figure 5. (**a**) The fitting results of NI/HTPE based on the KAS method, (**b**) the fitting results of NI/HTPE based on the FWO method, and (**c**) the activation energy calculation results of NI/HTPE based on the KAS and FWO methods.

3.2. Curing Kinetics Model and Parameters

Figure 6 shows the curves of dα/dt versus α for the N-100/HTPE, IPDI/HTPE, and NI/HTPE systems, all of which exhibit the characteristics of autocatalytic reactions [35,36]. In addition, some references also show that the reaction of a hydroxy-terminated polymer with isocyanate is autocatalytic [13,37–39]. Based on this, it is reasonable to assume that the reactions of these three systems are all autocatalytic reactions.

Equation (3) shows the autocatalytic reaction model [32], where m is the order of the autocatalytic reaction, n is the order of the non-catalytic reaction, and A is the pre-exponential factor. Equation (3) can be easily transformed to obtain Equation (4), and replacing α with $(1 - α)$ in Equation (4) gives Equation (5), while adding or subtracting Equation (4) from Equation (5) gives Equations (6) and (7) [40]. From Equations (6) and (7), it can be seen that by linearly fitting the scatter plot of Value I to $\ln((1 - α)/α)$, the slope obtained is n − m. By linearly fitting the scatter plot of Value II to $\ln(α - α^2)$, the slope obtained is n + m, and the intercept is 2lnA. Further calculation can obtain the values of m, n, and A.

$$\frac{d\alpha}{dt} = A\exp\left(-\frac{E_a}{RT}\right)\alpha^m(1 - \alpha)^n \tag{3}$$

$$\ln\left(\frac{d\alpha}{dt}\right) = \ln A - \frac{E_a}{RT} + m\ln\alpha + n\ln(1 - \alpha) \tag{4}$$

$$\ln\left(\frac{\mathrm{d}(1-\alpha)}{\mathrm{d}t}\right) = \ln A - \frac{E_a}{RT'} + m\ln(1-\alpha) + n\ln\alpha \tag{5}$$

$$\text{Value I} = \ln\left(\frac{\mathrm{d}\alpha}{\mathrm{d}t}\right) + \frac{E_a}{RT} - \ln\left[\frac{\mathrm{d}(1-\alpha)}{\mathrm{d}t}\right] - \frac{E_a}{RT'} = (n-m)\ln\left(\frac{1-\alpha}{\alpha}\right) \tag{6}$$

$$\text{Value II} = \ln\left(\frac{\mathrm{d}\alpha}{\mathrm{d}t}\right) + \frac{E_a}{RT} + \ln\left[\frac{\mathrm{d}(1-\alpha)}{\mathrm{d}t}\right] + \frac{E_a}{RT'} = (n+m)\ln\left(\alpha-\alpha^2\right) + 2\ln A \tag{7}$$

Assuming that the curing reactions of N-100/HTPE, IPDI/HTPE, and NI/HTPE follow the autocatalytic reaction model, according to Equations (6) and (7), it can be seen that Value I and Value II, obtained by the mathematical processing of the experimental data of the three curing systems, should have good linear relationships with $\ln((1-\alpha)/\alpha)$ and $\ln(\alpha-\alpha^2)$, respectively. This is a method that can be used to verify whether the curing reaction conforms to the autocatalytic reaction model. Based on this, the experimental data of the three curing systems were mathematically processed according to Equations (6) and (7) and plotted and linearly fitted. The results are shown in Figures 7–9.

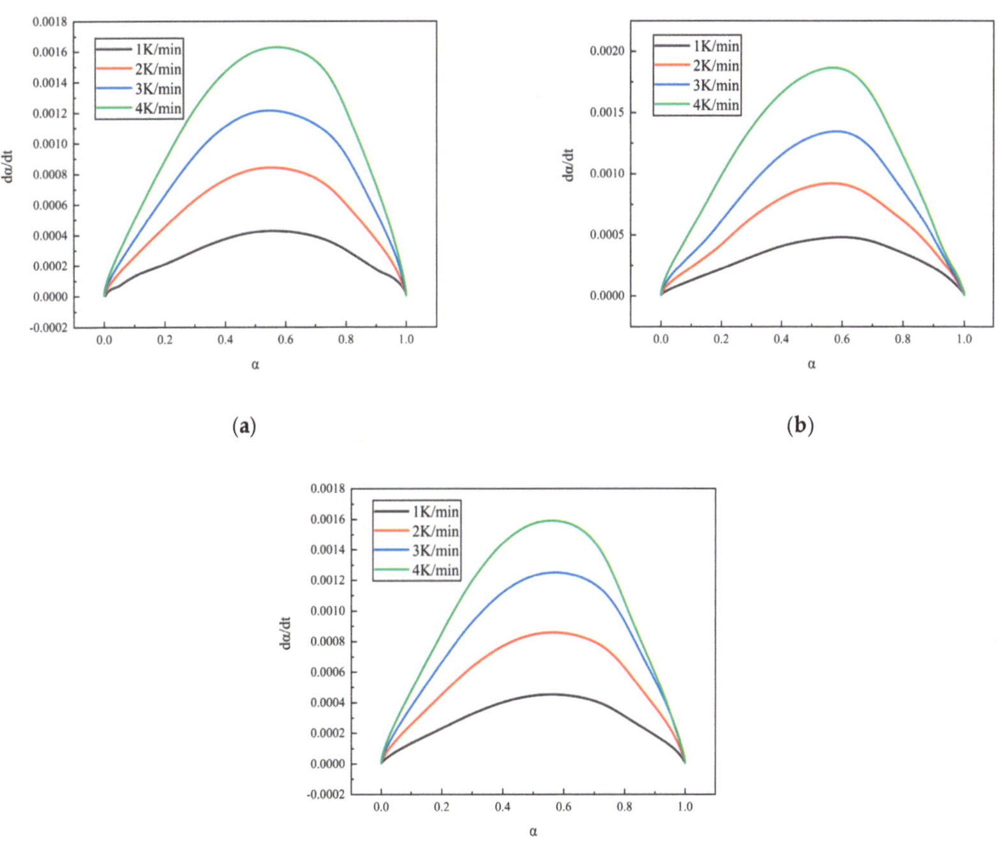

Figure 6. (**a**) The reaction rate $\mathrm{d}\alpha/\mathrm{d}t$ versus α curves of N-100/HTPE, (**b**) the reaction rate $\mathrm{d}\alpha/\mathrm{d}t$ versus α curves of IPDI/HTPE, and (**c**) the reaction rate $\mathrm{d}\alpha/\mathrm{d}t$ versus α curves of NI/HTPE. The colors of the curves in each graph represent the heating rates, with black, red, blue, and green representing heating rates of 1 K/min, 2 K/min, 3 K/min, and 4/min, respectively.

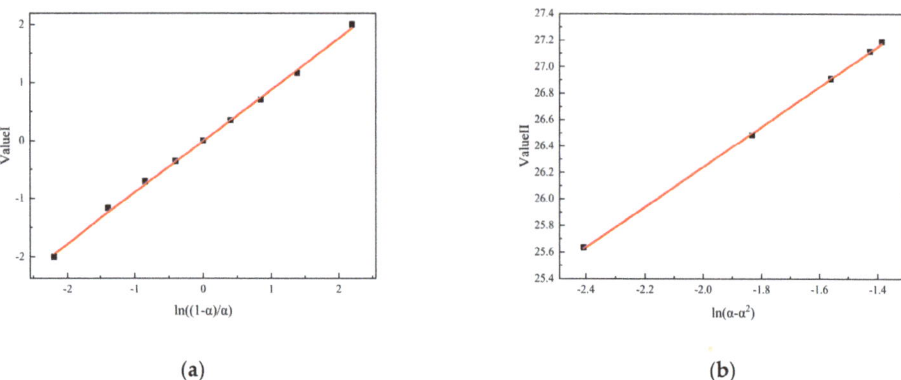

(a)   (b)

Figure 7. (**a**) The typical plots of Value I calculated using DSC data for N-100/HTPE and (**b**) the typical plots of Value II calculated using DSC data for N-100/HTPE.

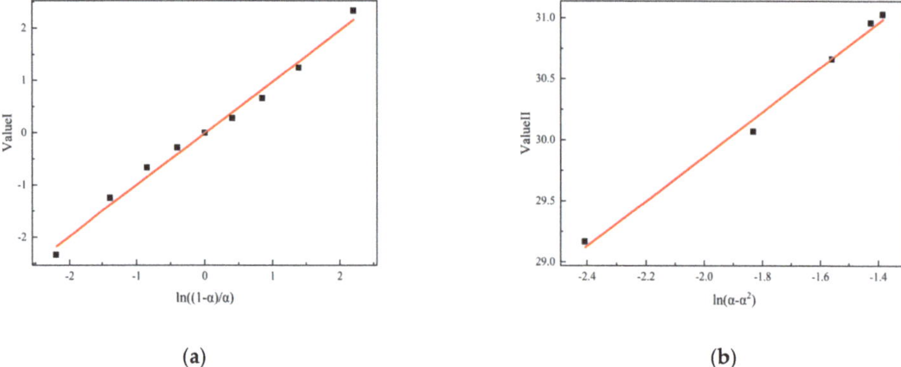

(a)   (b)

Figure 8. (**a**) The typical plots of Value I calculated using DSC data for IPDI/HTPE and (**b**) the typical plots of Value II calculated using DSC data for IPDI/HTPE.

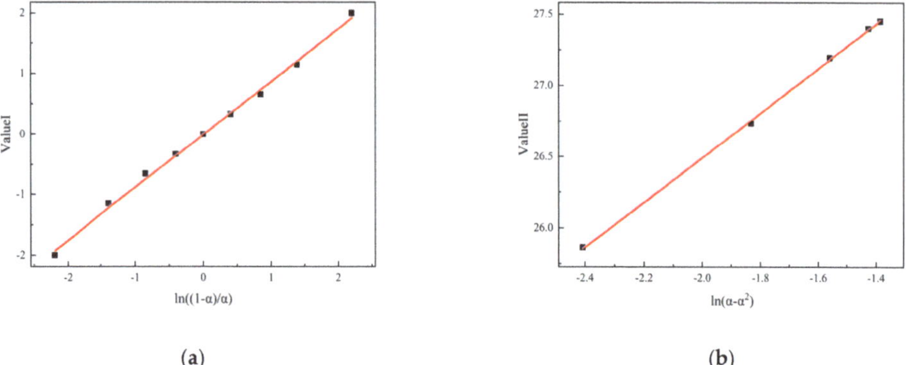

(a)   (b)

Figure 9. (**a**) The typical plots of Value I calculated using DSC data for NI/HTPE and (**b**) the typical plots of Value II calculated using DSC data for NI/HTPE.

From Figures 7–9, it can be seen that Value I and Value II of the three curing systems all have a relatively strong linear correlation with the corresponding $\ln((1 - \alpha)/\alpha)$ and

$\ln(\alpha - \alpha^2)$, indicating that the curing reactions of the three curing systems are all in line with the autocatalytic reaction model. Meanwhile, based on the linear fitting results, the corresponding kinetic parameters (m, n, and A) can be calculated and are listed in Table 2. On the other hand, after confirming that the reaction conforms to the autocatalytic reaction model, the experimental data were fed into Equation (3) and fitted using the Auto2Fit software (Professional Version 5.5) to obtain the corresponding kinetic parameters, which are also listed in Table 2.

Table 2. The curing kinetic parameters of the three curing systems obtained through calculation and fitting.

	Designation		N-100/HTPE	IPDI/HTPE	NI/HTPE
m	Kissinger	calculated	0.314	0.421	0.347
		fitting	0.274	0.456	0.311
	Ozawa	calculated	0.282	0.391	0.315
		fitting	0.242	0.428	0.279
	KAS	calculated	0.325	0.506	0.358
		fitting	0.284	0.535	0.322
	FWO	calculated	0.293	0.472	0.326
		fitting	0.252	0.504	0.290
n	Kissinger	calculated	1.200	1.407	1.220
		fitting	1.156	1.409	1.176
	Ozawa	calculated	1.214	1.419	1.234
		fitting	1.170	1.422	1.190
	KAS	calculated	1.195	1.373	1.215
		fitting	1.151	1.371	1.172
	FWO	calculated	1.209	1.386	1.229
		fitting	1.165	1.386	1.185
lnA	Kissinger	calculated	14.64	16.76	14.82
		fitting	14.57	16.79	14.75
	Ozawa	calculated	15.49	17.54	15.67
		fitting	15.42	17.57	15.61
	KAS	calculated	14.35	14.58	14.51
		fitting	14.29	14.60	14.44
	FWO	calculated	15.20	15.44	15.37
		fitting	15.14	15.46	15.30

From Table 2, it is evident that the kinetic parameters calculated using Equations (6) and (7) closely match the kinetic parameters obtained through the Auto2Fit software, suggesting the accuracy of the obtained kinetic parameters. Through comparison, it can be found that the autocatalytic order m of the three curing systems is smaller than the non-catalytic reaction order n, indicating that although the reactions of the three systems conform to the autocatalytic reaction model, the autocatalytic effect does not play a dominant role in the reaction process, which is also similar to related research [15,16,37,38,41]. This may partly explain why the activation energy increases as the conversion rate during the reaction process rises. Furthermore, it can be observed that the kinetic parameters of N-100/HTPE are lower than those of IPDI/HTPE, while the kinetic parameters of NI/HTPE are slightly higher than those of N-100/HTPE. Given that the curing agent used in the NI/HTPE system is a mixture of N-100 and IPDI, most of which is N-100, this phenomenon is understandable. In addition, it can be noted that the total reaction orders (m + n) of the three curing systems are all greater than 1 and less than 2, which is also normal because the total reaction orders of almost all polyurethane curing reactions are within this range of 1 to 2, indicating that the calculated reaction order parameters are reasonable.

3.3. Model Validation

Finally, by inputting the kinetic parameters from Table 2 into Equation (3), the curve of the dα/dt versus temperature for each curing system can be obtained. By comparing it with the experimental data, the accuracy of the reaction model and kinetic parameters can be verified. The results are shown in Figures 10–12, from which it can be seen that the calculated results and the fitted results agree well with the experimental curve, proving the validity of the reaction kinetics model and related parameters.

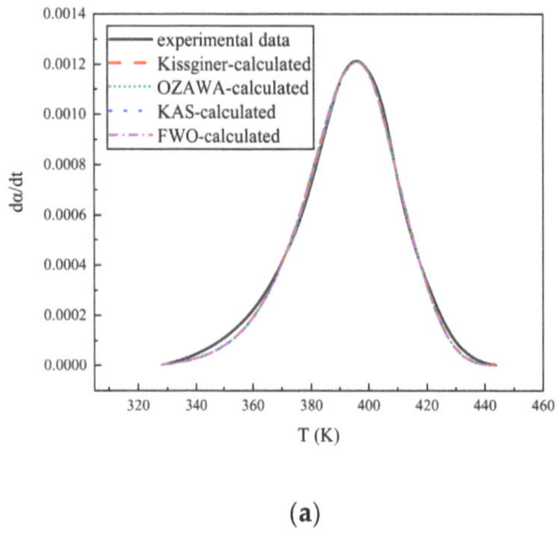

(**a**)

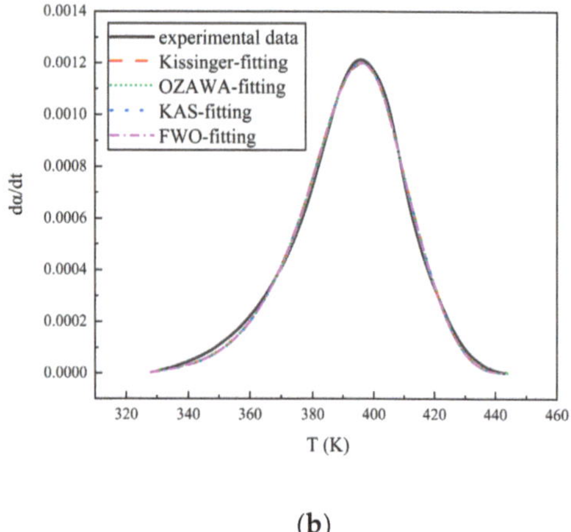

(**b**)

Figure 10. (**a**) The comparison of the experimental data of N-100/HTPE with the calculated results and (**b**) the comparison of the experimental data of N-100/HTPE with the fitting results.

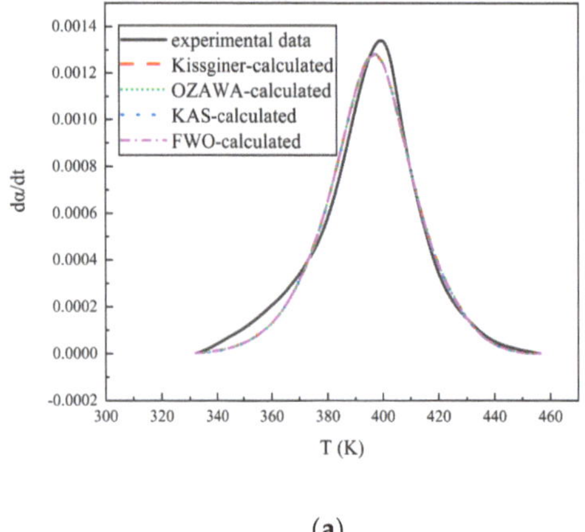

(**a**)

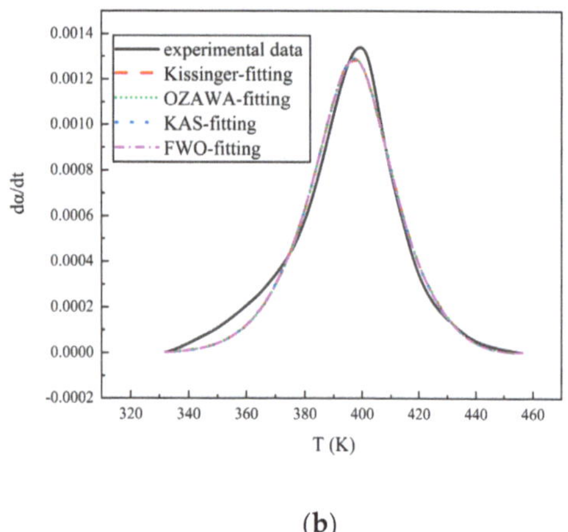

(**b**)

Figure 11. (**a**) The comparison of the experimental data of IPDI/HTPE with the calculated results and (**b**) the comparison of the experimental data of IPDI/HTPE with the fitting results.

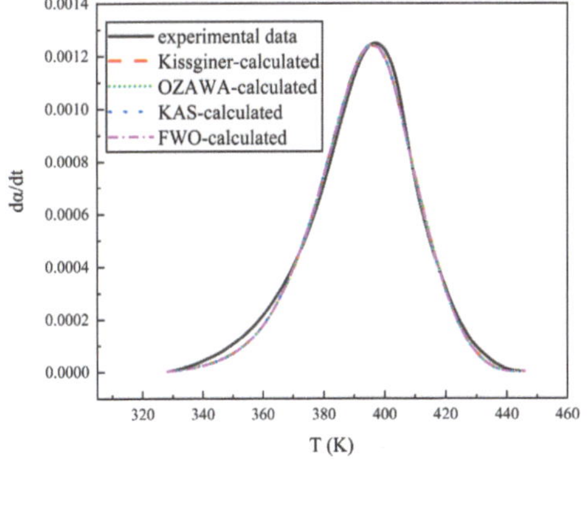

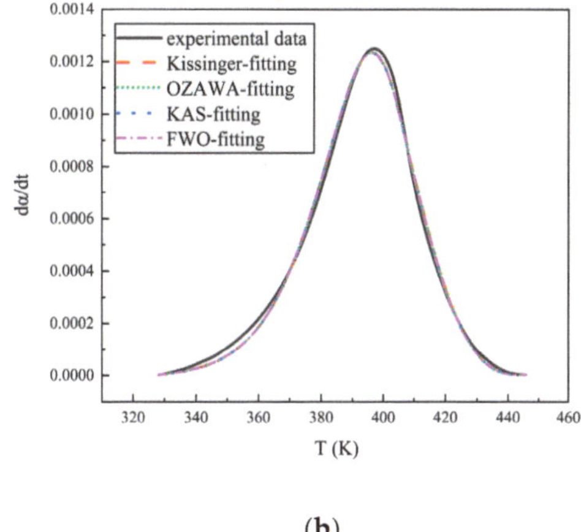

(a)   (b)

Figure 12. (a) The comparison of the experimental data of NI/HTPE with the calculated results and (b) the comparison of the experimental data of NI/HTPE with the fitting results.

4. Conclusions

The activation energy of the reaction between N-100 and HTPE is 69.37 kJ/mol, slightly lower than that of the reaction between IPDI and HTPE (75.60 kJ/mol), and the curing activation energy of the mixed curing agent and HTPE is 69.79 kJ/mol. However, when N-100 and IPDI are mixed as curing agents, both curing agents should react with HTPE at the same time, without a clear sequence. In addition, it has been proven through both calculation and fitting that the curing reactions of the N-100/HTPE, IPDI/HTPE, and NI/HTPE systems all conform to the autocatalytic reaction model. The non-catalytic reaction order n of N-100/HTPE is about 1.2, and the autocatalytic order m is about 0.3, both of which are lower than those of IPDI/HTPE, while the non-catalytic reaction order and autocatalytic order of NI/HTPE are between those of N-100/HTPE and IPDI/HTPE. Model validation shows that the obtained model and corresponding parameters have high accuracy. The research results of the curing kinetics obtained in this work contribute to a better understanding of the curing process of HTPE and provide guidance for its practical application.

Author Contributions: Conceptualization, Y.L., W.Y. and T.L.; methodology, W.Y.; software, W.L.; validation, W.Y., C.Z. and W.L.; formal analysis, T.L.; investigation, T.L.; resources, Z.D. and Y.L.; data curation, C.Z.; writing—original draft preparation, W.Y. and Z.D.; writing—review and editing, W.Y. and Y.L.; visualization, C.Z.; supervision, Z.D.; project administration, Y.L.; All authors have read and agreed to the published version of the manuscript.

Funding: This research received no external funding.

Institutional Review Board Statement: Not applicable.

Data Availability Statement: Data are contained within the article.

Conflicts of Interest: The authors declare no conflicts of interest.

References

1. Wu, W.; Jin, P.; Zhao, S.; Luo, Y. Mechanism of AP effect on slow cook-off response of HTPE propellant. *Thermochim. Acta* **2022**, *715*, 179291. [CrossRef]
2. Shi, L.; Fu, X.; Li, Y.; Wu, S.; Meng, S.; Wang, J. Molecular Dynamic Simulations and Experiments Study on the Mechanical Properties of HTPE Binders. *Polymers* **2022**, *14*, 5491. [CrossRef] [PubMed]
3. Zhang, H.-N.; Chang, H.; Li, J.-Q.; Li, X.-J.; Wang, H. High-strain-rate mechanical response of HTPE propellant under SHPB impact loading. *AIP Adv.* **2021**, *11*, 035145. [CrossRef]

4. Yuan, S.; Zhang, B.; Wen, X.; Chen, K.; Jiang, S.; Luo, Y. Investigation on mechanical and thermal properties of HTPE/PCL propellant for wide temperature range use. *J. Therm. Anal. Calorim.* **2021**, *147*, 4971–4982. [CrossRef]

5. Qian, Y.; Wang, Z.; Chen, L.; Liu, P.; Jia, L.; Dong, B.; Li, H.; Xu, S. A study on the decomposition pathways of HTPB and HTPE pyrolysis by mass spectrometric analysis. *J. Anal. Appl. Pyrolysis* **2023**, *170*, 105929. [CrossRef]

6. Shen, C.; Yan, S.; Ou, Y.; Jiao, Q. Influence of Fluorinated Polyurethane Binder on the Agglomeration Behaviors of Aluminized Propellants. *Polymers* **2022**, *14*, 1124. [CrossRef]

7. Shen, C.; Yan, S.; Tan, Y.; Ou, Y.; Jiao, Q.; Luo, Y. Enhancing energy release of aluminized propellants and explosives through fluorinated binder. *Propellants Explos. Pyrotech.* **2024**, *49*, e202300199. [CrossRef]

8. Yao, Q.; Xia, M.; Wang, C.; Yang, F.; Yang, W.; Luo, Y. A new fluorocarbon adhesive: Inhibiting agglomeration during com-bustion of propellant via efficient F–Al2O3 preignition reaction. *Carbon Energy* **2024**, *6*, e467. [CrossRef]

9. Xu, S.; Pang, A.; Kong, J. Decomposition mechanism and kinetic investigation of novel urea burning rate suppressants in AP/HTPE propellants. *J. Therm. Anal. Calorim.* **2023**, *148*, 12811–12820. [CrossRef]

10. Wu, W.; Zhang, X.; Jin, P.; Zhao, S.; Luo, Y. Mechanism of PSAN effect on slow cook-off response of HTPE propellant. *J. Energetic Mater.* **2022**, *42*, 313–330. [CrossRef]

11. Nie, J.; Liang, J.; Zhang, H.; Zou, Y.; Jiao, Q.; Li, Y.; Guo, X.; Yan, S.; Zhu, Y. Evolution of structural damage of solid composite propellants under slow heating and effect on combustion characteristics. *J. Mater. Res. Technol.* **2023**, *25*, 5021–5037. [CrossRef]

12. Lucio, B.; de la Fuente, J.L. Non-isothermal DSC and rheological curing of ferrocene-functionalized, hydroxyl-terminated polybutadiene polyurethane. *React. Funct. Polym.* **2016**, *107*, 60–68.

13. Lucio, B.; de la Fuente, J.L. Catalytic effects over formation of functional thermoplastic elastomers for rocket propellants. *Polym. Adv. Technol.* **2021**, *33*, 807–817. [CrossRef]

14. Ducruet, N.; Delmotte, L.; Schrodj, G.; Stankiewicz, F.; Desgardin, N.; Vallat, M.-F.; Haidar, B. Evaluation of hydroxyl ter-minated polybutadiene-isophorone diisocyanate gel formation during crosslinking process. *J. Appl. Polym. Sci.* **2013**, *128*, 436–443. [CrossRef]

15. Lucio, B.; de la Fuente, J.L. Rheological kinetics of ferrocenylsilane functionalized polyurethanes based on 4,4′-methylenediphenyl diisocyanate for advanced energetic materials. *J. Appl. Polym. Sci.* **2021**, *139*, 51765. [CrossRef]

16. Lucio, B.; de la Fuente, J.L. Kinetic and chemorheological modelling of the polymerization of 2,4- Toluenediisocyanate and ferrocene-functionalized hydroxyl-terminated polybutadiene. *Polymer* **2018**, *140*, 290–303. [CrossRef]

17. Duemichen, E.; Javdanitehran, M.; Erdmann, M.; Trappe, V.; Sturm, H.; Braun, U.; Ziegmann, G. Analyzing the network formation and curing kinetics of epoxy resins by in situ near-infrared measurements with variable heating rates. *Thermochim. Acta* **2015**, *616*, 49–60. [CrossRef]

18. Russo, C.; Serra, À.; Fernández-Francos, X.; De la Flor, S. Characterization of sequential dual-curing of thiol-acrylate-epoxy systems with controlled thermal properties. *Eur. Polym. J.* **2019**, *112*, 376–388. [CrossRef]

19. Magina, S.; Gama, N.; Carvalho, L.; Barros-Timmons, A.; Evtuguin, D.V. Lignosulfonate-Based Polyurethane Adhesives. *Materials* **2021**, *14*, 7072. [CrossRef]

20. Ding, S.; Fang, Z.; Yu, Z.; Wang, Q. Curing Kinetics and Mechanical Properties of Epoxy–Cyanate Ester Composite Films for Microelectronic Applications. *ACS Omega* **2023**, *8*, 32907–32916. [CrossRef]

21. Jouyandeh, M.; Yarahmadi, E.; Didehban, K.; Ghiyasi, S.; Paran, S.M.R.; Puglia, D.; Ali, J.A.; Jannesari, A.; Saeb, M.R.; Ranjbar, Z.; et al. Cure kinetics of epoxy/graphene oxide (GO) nanocomposites: Effect of starch functionalization of GO nanosheets. *Prog. Org. Coat.* **2019**, *136*, 105217. [CrossRef]

22. Vyazovkin, S.; Achilias, D.; Fernandez-Francos, X.; Galukhin, A.; Sbirrazzuoli, N. ICTAC Kinetics Committee recommenda-tions for analysis of thermal polymerization kinetics. *Thermochim. Acta* **2022**, *714*, 179243. [CrossRef]

23. Ou, Y.; Sun, Y.; Guo, X.; Jiao, Q. Investigation on the thermal decomposition of hydroxyl terminated polyether based polyu-rethanes with inert and energetic plasticizers by DSC-TG-MS-FTIR. *J. Anal. Appl. Pyrol.* **2018**, *132*, 94–101. [CrossRef]

24. Wen, X.; Zhang, G.; Chen, K.; Yuan, S.; Luo, Y. Enhancing the Performance of an HTPE Binder by Adding a Novel Hyper-branched Multi-Arm Azide Copolyether. *Propellants Explos. Pyrotech.* **2020**, *45*, 1065–1076. [CrossRef]

25. Chen, K.; Wen, X.; Li, G.; Pang, S.; Luo, Y. Improvement of mechanical properties of in situ-prepared HTPE binder in pro-pellants. *RSC Adv.* **2020**, *10*, 30150–30161. [CrossRef]

26. Attaei, M.; Loureiro, M.V.; Vale, M.D.; Condeço, J.A.D.; Pinho, I.; Bordado, J.C.; Marques, A.C. Isophorone Diisocyanate (IPDI) Microencapsulation for Mono-Component Adhesives: Effect of the Active H and NCO Sources. *Polymers* **2018**, *10*, 825. [CrossRef]

27. Wang, G.; Li, K.; Zou, W.; Hu, A.; Hu, C.; Zhu, Y.; Chen, C.; Guo, G.; Yang, A.; Drumright, R.; et al. Synthesis of HDI/IPDI hybrid isocyanurate and its application in polyurethane coating. *Prog. Org. Coat.* **2015**, *78*, 225–233. [CrossRef]

28. Kim, C.K.; Bae, S.B.; Ahn, J.R.; Chung, I.J. Structure-property relationships of hydroxy-terminated polyether based polyu-rethane network. *Polym. Bull.* **2008**, *61*, 225–233. [CrossRef]

29. Yuan, S.; Jiang, S.; Luo, Y. Cross-linking network structures and mechanical properties of novel HTPE/PCL binder for solid propellant. *Polym. Bull.* **2020**, *78*, 313–334. [CrossRef]

30. Balakrishnan, R.; Nallaperumal, A.M.; Manu, S.K.P.; Varghese, L.A.; Sekkar, V. DSC assisted kinetic analysis on the urethane network formation between castor oil based ester polyol and poly(methylene di phenyl isocyanate) (pMDI). *Int. J. Polym. Anal. Charact.* **2022**, *27*, 132–146. [CrossRef]

31. Zvetkov, V.; Simeonova-Ivanova, E.; Calado, V. Comparative DSC kinetics of the reaction of DGEBA with aromatic diamines IV. Iso-conversional kinetic analysis. *Thermochim. Acta* **2014**, *596*, 42–48. [CrossRef]

32. Olejnik, A.; Gosz, K.; Piszczyk, Ł. Kinetics of cross-linking processes of fast-curing polyurethane system. *Thermochim. Acta* **2020**, *683*, 178435. [CrossRef]

33. Park, S.; Yang, L.; Park, B.-D.; Du, G. Thermal cure kinetics of highly branched polyurea resins at different molar ratios. *J. Therm. Anal. Calorim.* **2023**, *148*, 6389–6405. [CrossRef]

34. Papadopoulos, E.; Ginic-Markovic, M.; Clarke, S. A thermal and rheological investigation during the complex cure of a two-component thermoset polyurethane. *J. Appl. Polym. Sci.* **2009**, *114*, 3802–3810. [CrossRef]

35. Zvetkov, V.; Calado, V. Comparative DSC kinetics of the reaction of DGEBA with aromatic diamines. III. Formal kinetic study of the reaction of DGEBA with diamino diphenyl methane. *Thermochim. Acta* **2013**, *560*, 95–103. [CrossRef]

36. Tezel, G.B.; Sarmah, A.; Desai, S.; Vashisth, A.; Green, M.J. Kinetics of carbon nanotube-loaded epoxy curing: Rheometry, differential scanning calorimetry, and radio frequency heating. *Carbon* **2021**, *175*, 1–10. [CrossRef]

37. Lee, S.; Choi, C.H.; Hong, I.-K.; Lee, J.W. Polyurethane curing kinetics for polymer bonded explosives: HTPB/IPDI binder. *Korean J. Chem. Eng.* **2015**, *32*, 1701–1706. [CrossRef]

38. Lucio, B.; de la Fuente, J.L. Chemorheology and Kinetics of High-Performance Polyurethane Binders Based on HMDI. *Macromol. Mater. Eng.* **2021**, *306*, 2000617. [CrossRef]

39. Lucio, B.; de la Fuente, J.L. Structural and thermal degradation properties of novel metallocene-polyurethanes. *Polym. Degrad. Stab.* **2017**, *136*, 39–47. [CrossRef]

40. Saeb, M.R.; Rastin, H.; Nonahal, M.; Ghaffari, M.; Jannesari, A.; Formela, K. Cure kinetics of epoxy/MWCNTs nanocomposites: Nonisothermal calorimetric and rheokinetic techniques. *J. Appl. Polym. Sci.* **2017**, *134*, 45221. [CrossRef]

41. Lucio, B.; de la Fuente, J.L. Kinetic and thermodynamic analysis of the polymerization of polyurethanes by a rheological method. *Thermochim. Acta* **2016**, *625*, 28–35. [CrossRef]

Article

Thermodynamic Coupling Forming Performance of Short Fiber-Reinforced PEEK by Additive Manufacturing

Qili Sun [1,*,†], Xiaomu Wen [2,*,†], Guangzhong Yin [3], Zijian Jia [2] and Xiaomei Yang [4]

[1] College of Mechanical and Electrical Engineering, Nanjing University of Aeronautics and Astronautics, Nanjing 210016, China
[2] National Key Laboratory of Transient Impact, No.208 Research Institute of China Ordnance Industries, Beijing 102202, China; zjjr116@163.com
[3] Escuela Politécnica Superior, Universidad Francisco de Vitoria, Ctra. Pozuelo-Majadahonda Km 1.800, 28223 Madrid, Spain; amos.guangzhong@ufv.es
[4] Fibre and Particle Engineering, University of Oulu, P.O. Box 4300, FI-90014 Oulu, Finland
* Correspondence: sunqili2023@nuaa.edu.cn (Q.S.); wenxm2908@163.com (X.W.)
† These authors contributed equally to this work.

Abstract: In this work, the PEEK/short carbon fiber (CF) composites were prepared, a new thermodynamic coupling (preheating and impact compaction) process of the FDM method is proposed, and the warp deformation mechanism was obtained by finite element simulation analysis. Results show that a new method could improve the forming quality of an FDM sample. The porosity of FDM samples of the PEEK/CF composite gradually decreased from 10.15% to 6.83% with the increase in impact temperature and frequency. However, the interlayer bonding performance was reduced from 16.9 MPa to 8.50 MPa, which was attributed to the influence of the printing layer height change from the printhead to the forming layer. To explain the above phenomenon, a thermodynamic coupling model was established and a relevant mechanism was analyzed to better understand the interlayer mechanical and porosity properties of PEEK/CF composites. The study reported here provides a reference for improving the forming quality of fabricated PEEK/CF composites by FDM.

Keywords: FDM; PEEK/CF composite; warp deformation; thermodynamic coupling; forming quality

Citation: Sun, Q.; Wen, X.; Yin, G.; Jia, Z.; Yang, X. Thermodynamic Coupling Forming Performance of Short Fiber-Reinforced PEEK by Additive Manufacturing. *Polymers* **2024**, *16*, 1789. https://doi.org/10.3390/polym16131789

Academic Editor: Chenggao Li

Received: 3 March 2024
Revised: 24 March 2024
Accepted: 29 March 2024
Published: 25 June 2024

1. Introduction

Polyether ether ketone (PEEK) is a semicrystalline thermoplastic polymer that is composed of repeating units containing one ketone bond and two ether bonds in its main chain structure [1]. Given its excellent comprehensive properties, such as wear resistance [2], fatigue resistance, and good biocompatibility [3], it has myriad applications in automotive and aerospace, medical equipment, and other industries [4]. Forming and manufacturing methods for fiber-reinforced PEEK-based composite materials are also attracting attention. PEEK has a glass transition temperature of 143 °C and a melting point of as high as 343 °C. Hence, the commonly used forming methods for this material are injection molding and compression molding. However, prototypes with complex structures still cannot be obtained through either approach. This situation has become a crucial problem that restricts the wide application of PEEK-based composites.

Therefore, many scholars have carried out research on FDM composite materials. Wang [5] systematically studied the influence of short carbon fibers (CFs) and orthogonal building orientation on the flexural properties of printed PEEK composites. They found that the addition of CF enhances the uniform nucleation of PEEK during 3D printing, decreases layer-to-layer bonding strength, and greatly changes the fracture mode. Li [6] reported the effects of the mechanical properties of PEEK printed through FDM and its composites on biocompatibility. Their experimental results confirmed that printed CFR-PEEK specimens have significantly improved mechanical properties compared with printed pure PEEK.

Laboratory experiments clearly showed that no toxic substances are introduced during the FDM manufacturing of pure PEEK and CFR-PEEK. Arif [7] studied the multifunctional performance of PEEK composites reinforced with carbon nanotubes (CNTs) and graphene nanoplatelets (GNPs). They reported that the Young's and storage moduli of these materials increased by 20% and 66% under 3 wt.% CNT loading, respectively, and by 23% and 72% under 5 wt.% GNP loading, respectively. Moreover, they demonstrated that the crystallization temperature and crystallinity degree of FFF-PEEK increase with the addition of carbon nanostructures [8]. Tian [9] proposed three methods to improve the interface performance. Firstly, to improve the wettability of the resin and the fiber, the solution impregnation technology was used to pretreat the carbon fiber, and a PA slurry layer was introduced onto the surface of the carbon fiber to improve the fiber surface and resin interface bonding performance [10]. Secondly, physical cleaning and chemical modification of the carbon fiber surface by plasma treatment can effectively improve the bonding properties of two materials with a large polarity difference [11]. The interlayer performance of the 3D-printed sample after plasma treatment is improved by 70%. The effect of laser power and printing speed on the extrusion of continuous carbon fiber-reinforced PEEK composites is as follows: with the increase in laser power, the temperature of the junction point increases rapidly, the temperature of the binding point reaches about 420 °C, and the thermal accumulation of the surface layer decreases with the increase in the printing speed; 120 mm/min is the fastest printing speed at the highest thermal accumulation temperature, the interlaminar property of PEEK/CF can reach 56 MPa [12]. Thirdly, the interlaminar bonding property of PEEK/CF can obviously be improved with the decrease in scanning distance and interlaminar thickness. When the fiber content reaches about 35 wt.%, the interlaminar shear strength can reach about 35 MPa [11].

However, most of the studies focused on the printing and forming mechanical properties of 3D-printed composites utilizing filaments [13], and few reported on the warp deformation and formed quality of PEEK-based composites fabricated through 3D printing. In the process of FDM printing, the temperature difference is large because the thermoplastic material changes from a molten state to a solid state, which easily causes stress concentration in the composite material system, leading to micro-cracks, interface delamination, warping deformation, and other defects of the composite sample. These seriously affect the quality and mechanical properties of the FDM composites [14,15]. At present, there are few literature reports on the influence of residual stress on the warping deformation mechanism of formed samples. In addition, in the forming process of FDM stacking layer by layer, each layer will be subjected to multiple high-temperature thermal cycles, resulting in a comprehensive superimposed heat effect, resulting in a temperature gradient in the space of the heated plane, resulting in the deformation of the formed parts due to uneven temperature, and further leading to problems such as difficult interlayer fusion and large differences in the mechanical properties of the composite materials. Therefore, it is urgent to study the new forming process of preheating-impact-printing in the FDM forming process to improve the physical fusion characteristics between the new forming layer and the old layer and to improve the forming quality of the sample.

Based on the above analysis, the warpage deformation of PEEK/CF composites were predicted by finite element analysis, and the warpage deformation factors were analyzed during the forming process. Then, a new forming process integrating preheating, impact, and printing was proposed, and the influence of this process on the mechanics and forming quality of the forming samples was studied, which provides an important reference for improving the performance of FDM printing samples.

2. Experimental Materials and Test Methods

2.1. Experimental Materials

PEEK has a glass transition temperature of 143 °C and a melting point of as high as 343 °C. PEEK (450 G) was supplied by VICTREX Company (Lancashire, UK). Chopped CF (WD-100 AW) with an average fiber length of 100 μm was purchased from Nanjing Weida

Composite Material Co., Ltd. (Nanjing, China) Prior to use, the PEEK and CF were dried for at least 3 h at 150 °C.

PEEK and PEEK/CF composite filaments were supplied by Shanxi Jugao AM Co., Ltd. (Taiyuan, China) The CF contents of the PEEK/CF composites were approximately 10 wt.%. The diameters of the filaments for 3D printing were controlled to 1.75 ± 0.05 mm.

2.2. Test Equipment and Test Conditions

(1)   Shear strength test

Instrument: WDW-1 universal material testing machine, Beijing University of Chemical Technology Testing and Analysis Center.

Test conditions: Interlayer shear strength analysis was performed in reference to the standard ISO 14130:1998 (fiber-reinforced plastic composite material short beam method to determine the interlayer shear strength). The sample size was 20 mm × 10 mm × 3 mm, and the loading speed was 5 mm/min. The span was 10 mm, the upper support radius was 5 mm, and the lower support radius was 2 mm. The formula used to calculate the interlaminar shear strength τ_M (MPa) is:

$$\tau_M = \frac{3F}{4bh}$$

where F is the maximum load (N), b is the sample width (mm), and h is the sample thickness (mm).

(2)   Porosity test (mercury intrusion method)

Instrument: POREMASTER GT60 mercury prosimeter. Test data were obtained by the Beijing Center for Physical and Chemical Analysis.

Test conditions: room temperature, pressure range of 0.20–30,000 psi, and mercury injection time of 70 min.

2.3. Fabrication of 3D-Printed Specimens

Five samples of each geometry were created by using PEEK/CF material to investigate the relationships among printing factors, mechanical properties, and porosity. The printing process is shown in Figure 1. The specifications of the geometric models of shear stress, GB/T1450.1-2005. The test sample models conforming to the relevant test standards were designed in Solidwork software 2020, and the geometric models were exported as files in stereolithography format for importation by the FDM software (Simplify 3D 3.0). The main FDM process parameters used to print the PEEK/CF composite samples are provided in Table 1.

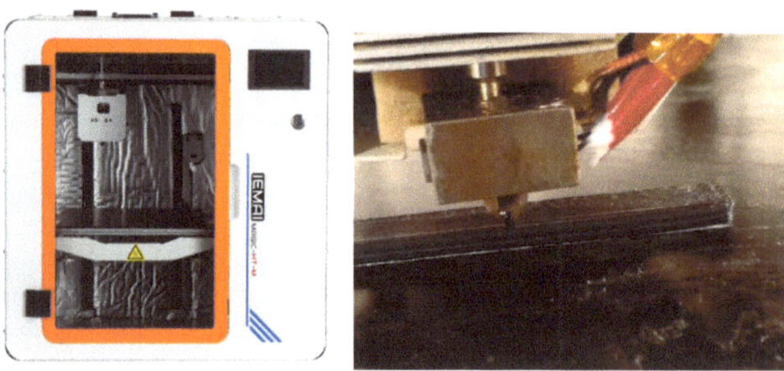

Figure 1. IEMAI 3D equipment and print sample process by FDM.

Table 1. Printing process parameters.

Parameter	Value
Print head diameter/mm	0.4
Print head temperature/°C	400
Base plate temperature/°C	100
Print speed/(mm/min)	500
Print width/mm	0.5
Layer thickness/mm	0.3

2.4. Design of Multifunction Printhead

The multifunctional printhead is designed as a "fly swatter" structure and surrounds the printhead, as shown in Figure 2. The multifunction printhead includes the body printhead, impact compaction device, electric motor, and electromagnetic pulse controller. The device adopts an electric heating rod with a parallel structure and the corresponding temperature sensor, and the temperature of the electric heating rod is controlled by a micro-temperature controller to ensure the preheating effect of the impact device and uniform and stable preheating. The impact compaction device is used for the impact force size adjustment; an external digital dynamometer is also used for testing in the range of 0–5 N; the dynamometer is connected to the detection computer; the computer is utilized to operate the digital dynamometer through the software interface; and the test data are observed.

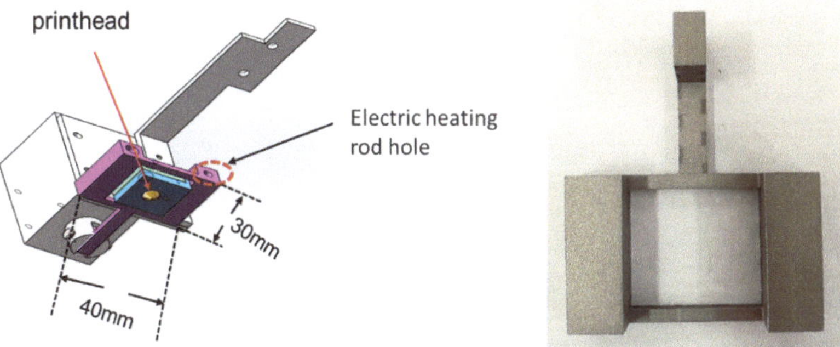

Figure 2. Preheating and shock device.

3. Analysis and Discussion of Experimental Results

3.1. Finite Element Simulation Analysis of Forming Warpage Deformation

3.1.1. Hypothesis of Finite Element Model Analysis of FDM Process

The FDM process is a process in which PEEK wire is fused and extruded by a nozzle. During the printing process, the nozzle moves on the surface of the workpiece, the temperature gradient between the wire and the environment is large, and the heat source is concentrated. As a result, the temperature field is unsteady, and heat conduction and convection are the main heat transfer methods.

According to the abovementioned forming process characteristics, the numerical simulation model is established under the following assumptions:

(1) The material continuity assumption is that the material is treated with continuous heat transfer throughout the forming process.

(2) The model adopts a 3D finite element model for transient thermal analysis because the temperature field is unsteady.

(3) According to the printing process of a specific path, the "birth and death unit" method can be used for simulation, which activates regional units and loads heat sources one by one, according to the scanning speed of the device.

(4) The process of FDM has a problem of latent heat-of-phase transition in material forming, which is treated by the equivalent thermal melting method.

(5) The initial temperature of the unit is the temperature of the nozzle extrusion, and the convection heat transfer is conducted with the air after extrusion.

3.1.2. Heat Source Model Analysis

According to the molding process of the 3D printer, the silk is heated by the nozzle from a solid state to a molten state. The nozzle is printed according to a specific path. Thus, the nozzle can be regarded as a moving heat source.

At present, heat source models can mainly be divided into the double-ellipsoid heat source model and the Gaussian heat source model. For the preparation process of molten deposition, the heat of the nozzle in the accumulation molding process is relatively concentrated, but no penetration effect occurs. The same position will also have differences with the change in time, and a large temperature gradient change will occur in the spatial position. This characteristic corresponds to the Gaussian heat source model. Thus, the Gaussian heat source model is used to simulate the nozzle heat source.

The heat source model of the Gaussian function distribution is shown in Figure 3 below:

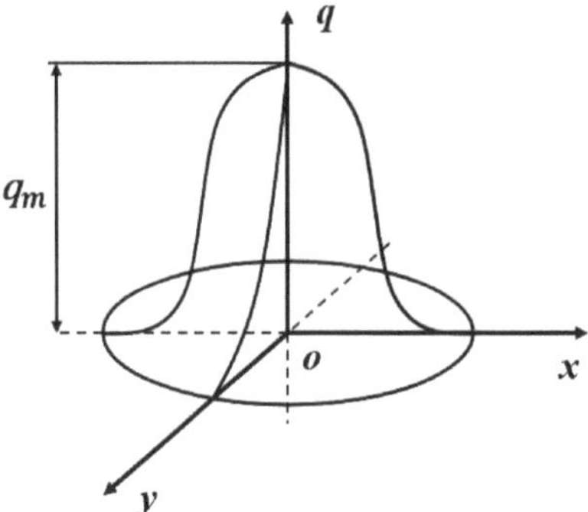

Figure 3. Heat flux distribution of the Gaussian heat source [16].

The heat flux density distribution on the heating spot adopts the Gaussian mathematical model, and its heat flux density distribution function is as follows [17]:

$$q(r) = \frac{3\eta P}{\pi R^2} \times \exp\left(-\frac{3r^2}{R^2}\right) \tag{1}$$

where R is the effective heating radius of the heat source, r is the distance from any point on the heating surface to the center of the arc heating half, η is the effective power coefficient that truly acts on the molded parts after environmental effects such as convection and heat exchange, and P is the rated power of the heat source.

According to the discussion above, the heat of the nozzle is relatively concentrated, the heat source of the printing process is the moving Gaussian surface heat source, and its moving trajectory is the printing path. Figure 4 shows the trajectory of heat source movement.

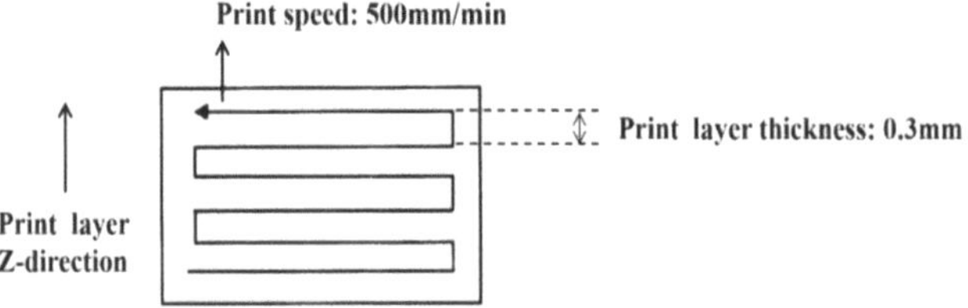

Figure 4. Trajectory of heat source movement.

3.1.3. Simulation Analysis of Warpage Deformation of Printed Samples

(1) Finite element model

① Model of pure PEEK composite

The model uses the printing unit microsegment as the analysis object, and the structure is shown in Figure 5. The model length is 0.5 mm × 0.2 mm × 0.1 mm, and it contains 87,567 nodes and 81,400 units.

Figure 5. Model of pure PEEK composite.

② Model of fiber-reinforced PEEK composite

PEEK/short carbon fiber (CF) is a composite material reinforced with 10% chopped carbon fiber with a mixed volume fraction. The staple fiber is a strong anisotropic material, and the staple fiber is randomly mixed in the PEEK material. The model selects to print 5 mm microsegments as the analysis object. First, the discrete random algorithm is applied to discretize the short fibers into the model. Then, each short fiber is given random spatial direction in the model. Finally, the model is integrated for analysis. The model is shown in Figure 6.

Figure 6. Model of PEEK/CF composite.

(2) Simulation of printing layer sample

The five-layer sample is used as the analysis object to simulate the forming process of the multilayer sample; the heat source model is brought into the printing process simulation; the printing multilayer simulation method is like the single layer; the "life and death" unit technology is also used to simulate the printing process; and the unit "life and death" simulation path is consistent with the printing path. The heat flux density and temperature distribution cloud of the printed sample at a certain time are shown in Figure 7.

Figure 6 shows that the temperature of the sample during the printing process is the maximum near the printing terminus, and it diffuses radially to the periphery, which indicates a temperature gradient situation. During the printing process, the sample will absorb part of the heat due to the convection heat dissipation on the upper surface. As a result, the overall temperature of the sample tends toward the thickness direction of the ladder shape, and a trend of decreasing layer by layer is observed.

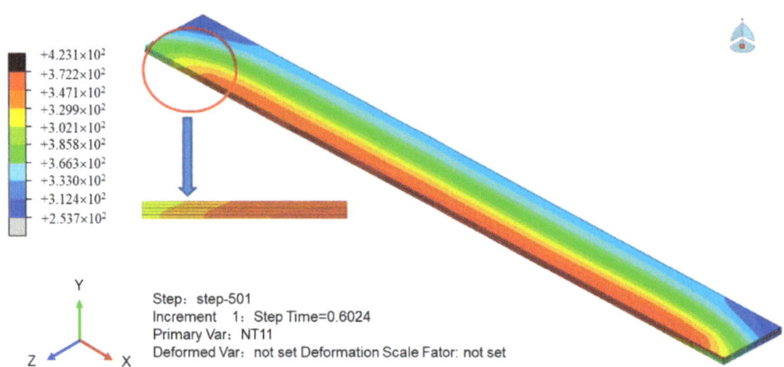

Figure 7. Temperature field cloud at the end of printing.

To obtain the relationship between the internal stress and temperature of the printed sample, we simulate the stress cloud and the displacement cloud in the thickness direction of the sample, as shown in Figures 8 and 9, respectively. Figure 8 shows that the stress of the sample near the printing end lane is minimal; it decreases to the perimeter of the printing termination, and the overall stress field is an inverse wave layer. As shown in Figure 9, the stress of the sample is affected by the temperature field. The temperature of the upper surface of the sample drops sharply, the stress gradually increases, and the upper part of the sample shrinks with the increase in time. Meanwhile, the middle and bottom surface of the sample have heat conduction due to the printing process, the temperature drops slowly, and the internal stress changes slowly with the temperature. In general, the warpage around the upper part of the molded part changes the most, and the deformation of the sample is concave.

The stress field value of the PEEK/CF composite is obtained by extracting the nodal displacement of the bottom surface in Figure 10, to obtain the intuitive warpage deformation characteristics of the sample. The figure shows that the fiber-reinforced PEEK material is scattered with the discreteness of the spatial position of the chopped fiber due to its mixing characteristics. The overall microsegment stress is uneven, and the interface stress between the fiber and PEEK material is greater and shows stress concentration. The two stress field distributions caused by material and temperature have different characteristics of bottom warpage, and the fiber-reinforced PEEK material with a large internal stress has a large amount of warpage deformation.

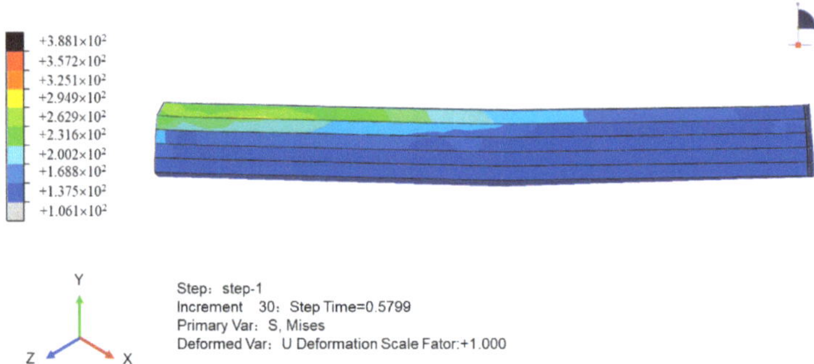

Figure 8. Sample stress cloud when printing is completed.

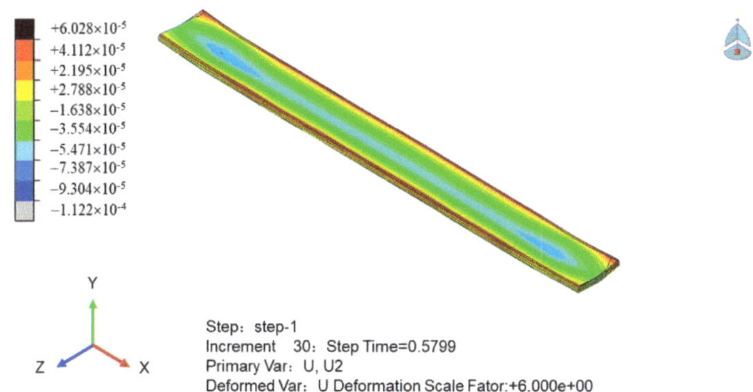

Figure 9. Pattern of sample thickness direction displacement cloud when printing is completed.

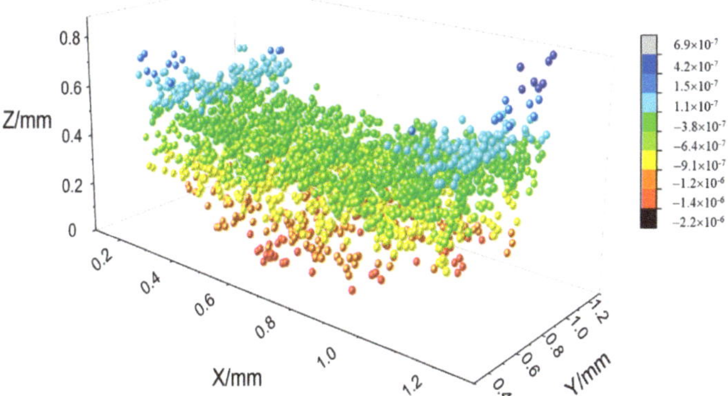

Figure 10. Calculation results of the stress field of PEEK/CF composites.

Figure 11 shows the actual printing situation, and the warpage deformation trend of the simulated printing results is consistent with the actual situation.

Figure 11. Printed samples (warpage deformation).

3.2. Integrated Forming Process of Preheating, Printing, and Impact Compaction

Previous studies have shown that the heat treatment temperature is higher than the glass transition temperature of PEEK thermoplastic material, PEEK shows high elastic characteristics, and the molecular chain segment has strong movement ability, which can effectively release the residual stress originally "sealing" in the matrix [18]. Therefore, the heat treatment process can reduce the maximum residual stress value of the fused deposition-forming sample, and the forming process of the PEEK/CF composite material needs to be heat treated. Therefore, we designed a multifunctional printhead to realize the forming process of preheating, printing, and impact compacting and expect to achieve the integrated forming of preheating, printing, and impact for improving the forming quality of the sample.

3.2.1. Interlaminar Mechanical Properties of PEEK/CF Composites

(1) Influence of preheating temperature on the mechanical properties of interlayers

In this section, the temperatures of the design hammer are 330, 340, 350, 360, and 370 °C, the impact frequency is 6 Hz, and the impact strength is 0.3 N. The influence of different preheating temperatures on the interlayer shear strength is studied, and the results are shown in Figure 12.

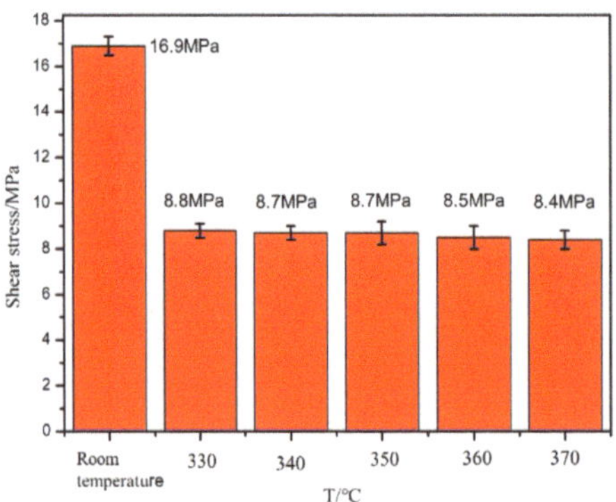

Figure 12. Effect of preheating temperature on shear strength.

Figure 12 shows that the interlayer shear strength of the blank sample is 16.9 MPa; the interlaminar shear strength of the impact sample is the same, with a value of approximately 8.5 MPa; and the interlaminar strength of the PEEK/CF composite sample prepared through FDM after impact is reduced. This result may be due to two reasons. One is that the thermal radiation temperature of the impact hammer is low. The impact effect is a short thermal contact, and the heat flow is transmitted to the resin surface for a short time, which causes the "physical cross-linking point" between the hanging molecular chains in the layers of PEEK thermoplastic materials to become smaller; hence, the physical cross-linking effect between the layers is weakened, which, in turn, leads to a decrease in the bonding strength between the layers. The other reason is that the layer height of the printed sample increases after the impact action. The height of the extruded resin layer is significantly reduced due to the impact effect, but the spacing from the printhead to the printing layer increases, and the preprinted layer height of the next layer is larger when the impact strength is higher. As observed from the printing background, the interlayer shear strength is smaller when the layer height is larger. The sample after the impact does not achieve the desired results because of these two reasons.

This study establishes a model for the influence of the height change from the printhead to the forming layer on the interlayer performance, as shown in Figure 13, to understand the reasons for the degradation of the interlayer performance of samples after impact. For the design printing process parameters, the printing layer height is h. After the impact, the layer height thickness is reduced Δh. When the next layer is printed, the originally designed layer height h now becomes $h_1(h + \Delta h)$. The printing layer height increases. Thus, the interlayer shear strength decreases.

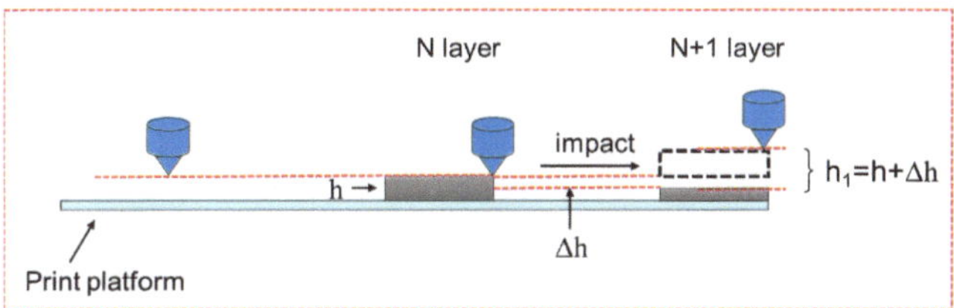

Figure 13. Chart of change in printing layer height.

A comparative experiment is designed in this section to study the influence of printing layer height on the interlayer performance of additive manufacturing of PEEK/CF composites for verifying the effect of the change in printing layer height on interlayer performance. Different layer heights of 0.1, 0.2, and 0.3 mm of FDM are designed, and the results are shown in Figure 14.

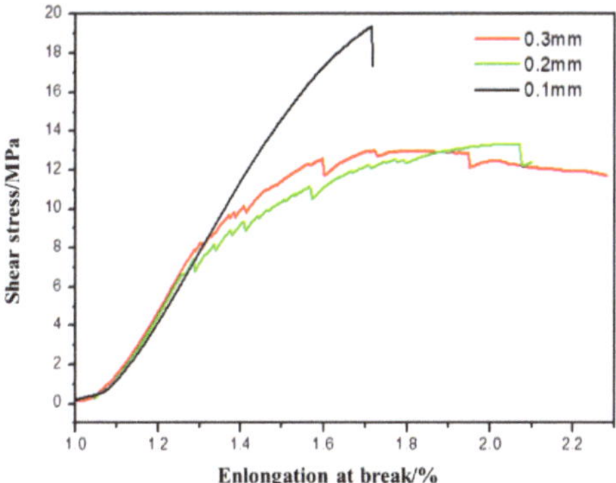

Figure 14. Effect of printing layer height on the shear strength of PEEK/CF composites.

The figure shows that the interlayer shear strength of PEEK/CF composite prepared through additive manufacturing gradually decreases with the increase in printing layer height. The interlayer shear strength of PEEK/CF composite prepared by FDM is 19.37 MPa when the layer height is 0.1 mm; however, its interlayer shear strength is greatly reduced to 11.20 MPa when the layer height is increased to 0.3 mm. The reason is that the printhead has a compacting function for the molded part, and the lower printing layer height can improve the interlayer performance. Therefore, the printing layer height is conducive to improving the interlayer performance of the formed sample. At the same time, ensuring the impact effect and appropriately designing the layer height compensation mechanism are particularly important for improving the interlayer performance.

(2) Influence of impact strength and frequency on interlayer mechanical properties

The real-time impact on the printing layer improves the quality of the formed sample. Preliminary studies have shown that the impact head will drive the printer to vibrate when the impact strength is higher than 1 N, and this condition is not conducive to printing. Therefore, the impact strength of this section is designed to be lower than 1 N (0, 0.3, 0.6, and 0.9 N), the preheating temperature and impact frequency are 350 °C and 6 Hz, respectively, and the influence of different impact strengths on the interlayer shear strength is studied, as shown in Figure 15, which shows that the interlayer shear strength of the blank sample is 16.9 MPa, and the interlayer shear strength of the sample after impact is lower than that of the blank sample. Compared with samples without any impact action, the impact strength also has a weakening effect on PEEK/CF composite samples prepared through FDM.

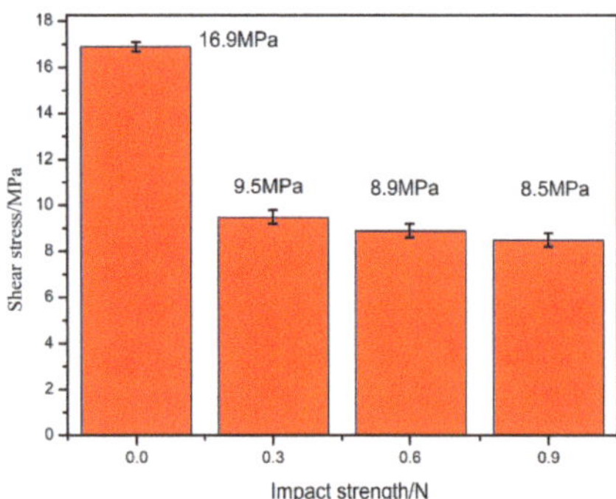

Figure 15. Effect of impact strength on interlaminar shear strength.

Different impact frequencies (0, 2, 4, and 6 Hz) and preheating temperatures and impact strengths of 350 °C and 0.3 N, respectively, are designed in this section to study the influence of different impact frequencies on the interlaminar shear strength, as shown in Figure 16, for exploring the influence of the impact frequency on the interlaminar performance. The figure shows that the interlayer shear strength of the blank sample is 16.9 MPa, and the interlayer shear strength of the sample after impact is lower than that of the blank sample.

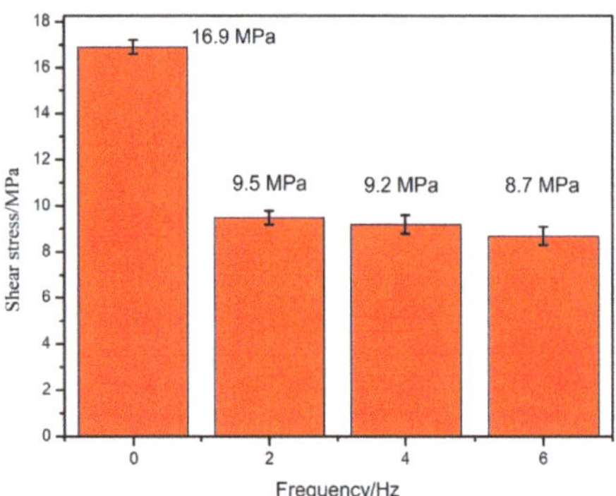

Figure 16. Effect of impact frequency on interlaminar shear strength.

For the cause analysis, compared with the sample without any impact effect, the impact frequency also has the effect of weakening the interlayer of the PEEK/CF composite samples prepared through FDM. The reason is still the increase in the height between layers and the low preheating contact temperature, which result in weak interlayer performance.

3.2.2. Porosity Analysis of Formed Samples under Thermodynamic Coupling

The porosity of formed samples under different preheating temperatures and impact frequencies, as shown in Figures 17 and 18, is studied in this section for investigating the porosity of FDM samples of PEEK/CF composites under thermodynamic coupling.

The influence of the preheating temperature on the porosity of PEEK/CF composites is studied when the design impact strength is 0.3 N and the impact frequency is 6 Hz. The influence of different preheating temperatures on the porosity of the formed samples is obtained. The results are shown in Figure 18. The figure shows that the porosity of the FDM samples of the PEEK/CF composite gradually decreases from 10.15% to 6.83% with the increase in preheating temperature. This result is due to the fact that the molecular segment movement of PEEK resin is more violent when the temperature is higher, and the gaps between the layers and the porosity of the material are lower when it is easier to compact under the action of an impact hammer. Therefore, preheating treatment during the printing process can reduce the porosity of the formed sample of the PEEK/CF composite.

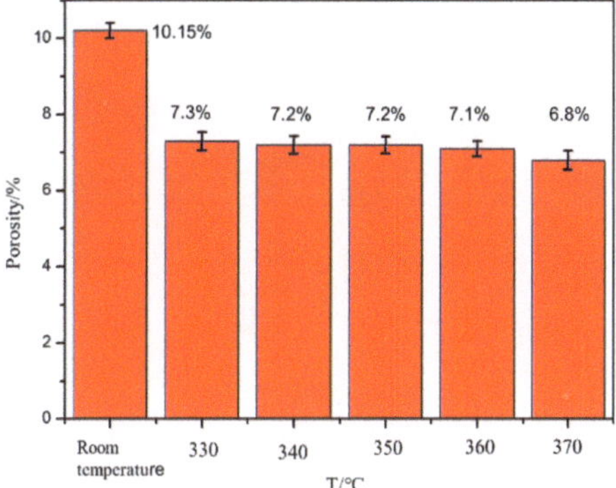

Figure 17. Effect of preheating temperature on porosity.

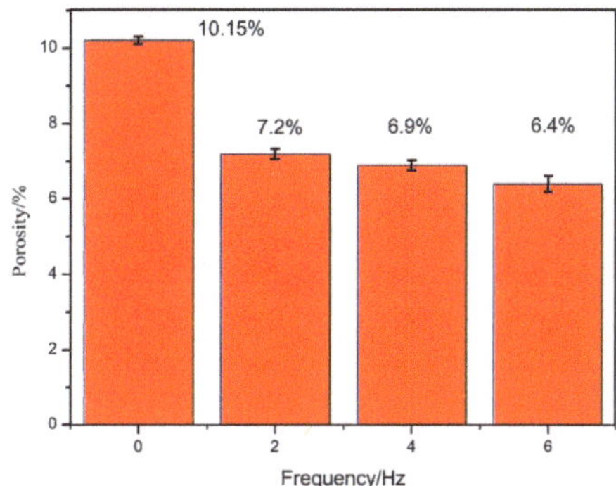

Figure 18. Effect of impact frequency on porosity.

The influence of different impact frequencies on porosity is studied with a preheating temperature of 350 °C and an impact strength of 0.3 N. The test results are shown in Figure 18. The figure shows that the porosity of the fused sedimentary samples of PEEK/CF composites (from 10.15% to 6.67%) gradually decreases with the increase in impact frequency. This result is due to the fact that the preheating temperature is constant, the PEEK/CF composite is easier to compact, and the porosity of its interlayer gaps and materials is lower when the impact hammer acts on the freshly printed sample more often. Therefore, the high impact frequency can reduce the porosity of the formed samples of PEEK/CF composites.

3.3. Mechanism of Thermodynamic Coupling

In the process of additive manufacturing of parts, the remelting process of the print nozzle is the consolidation process of rapid melting and solidification of resin, which contains complex phenomena affected by multiple physical factors, such as heat transfer, mass transfer, diffusion, and phase change. A thermodynamic coupling model of PEEK/CF by FDM is established and its thermodynamic coupling mechanism is analyzed to better understand the interlayer mechanical properties of PEEK/CF composites under thermodynamic coupling. The thermodynamic coupling mechanism is shown in Figure 19.

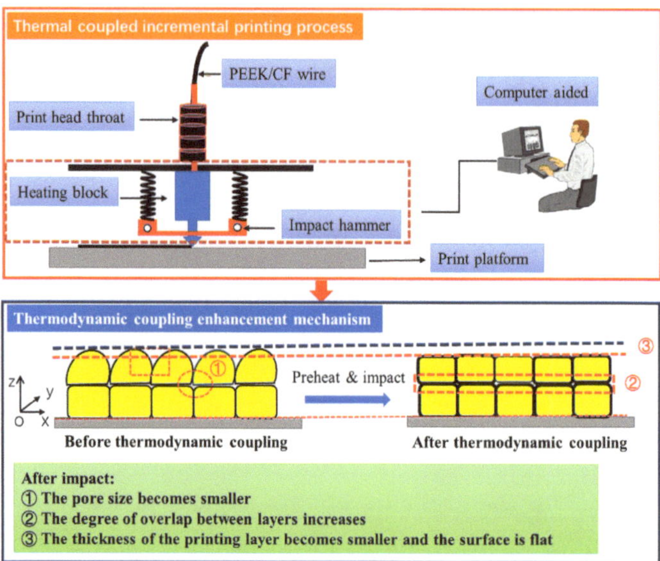

Figure 19. Thermodynamic coupling model.

FDM wire is a fiber-reinforced PEEK thermoplastic resin matrix composite material. In the forming process, the resin is melted, and after passing through the print head, it is fused, expanded, and then bonded onto the print layer. The macro-morphology shows a steamed bun shape, and the bond between layers is insufficient, resulting in higher interlayer porosity and a larger interlayer pore size. The high porosity ratio and large pore structure ultimately affect the interlayer mechanical properties. Therefore, in order to reduce the porosity and pore structure of the FDM forming process, a compact device similar to a "fly swatter" was designed, and a new forming process integrating preheating, printing, and compaction was proposed. The process is divided into three steps: the first is the preheating process, before printing the new forming layer, the forming layer to be covered, and preheating treatment (the temperature of the forming layer covered is generally about 100 °C printing chamber temperature, and the melting extrusion temperature of printing PEEK is about 340 °C), which can improve the melting bonding degree of the new printing layer and the old layer. The printing process is normally printed and overlaid on the "old layer". During the impact process, after the "old and new layers" to be printed are fused, the impact of the physical action on the new and old layers can better promote interlayer adhesion and reduce the porosity and pore size.

After preheating and impact, the FDM samples of PEEK/CF composites undergo three changes. First, the porosity of the composite material is reduced. Second, the overlap of interlayer bonding increases. Third, the thickness of the sample printing layer is reduced.

The system integration of a 3D printer based on a preheating and stamping device is completed, and a 3D printing experiment of carbon fiber-reinforced PEEK composite is conducted. The sample printing before and after preheating and shock is obtained as follows: Figure 20 shows that the sample without impact has a rough surface, and the surface has obvious ravines "plowed" by the printhead. After preheating and impact, the surface of the sample is relatively neat, which shows that the surface quality of the formed sample can be effectively improved by the preheating and impact process.

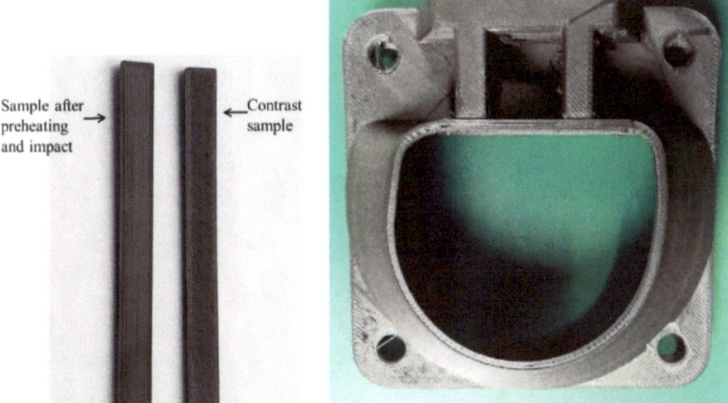

Figure 20. PEEK/CF composite before and after preheating and impact.

4. Conclusions

In this paper, PEEK/CF composites with 10% fiber content were prepared by FDM, and the following conclusions were obtained through comparative analysis and experimental testing:

(1) Finite element analysis shows that uneven heating and cooling and volume shrinkage caused by material phase change in the process of printing result in internal stress in the forming process of the printed sample, which leads to residual stress and deformation of the sample after forming. The fiber affects the stress field distribution of PEEK/CF composites and is affected by the material characteristics and temperature, and the warpage deformation is greater when the interface stress between the fiber and PEEK is greater.

(2) The printing verification of typical samples was conducted, which could effectively improve the printing performance and forming quality. The results show that the 3D-printed sample of carbon fiber-reinforced PEEK composite material printed without pressure has a rough surface. The surface of the sample has obvious ravines "plowed" by the printhead, and its surface is relatively neat after preheating and impact. The porosity study shows that the preheating and impact effects can reduce the porosity of the formed sample from 10.15% to 6.83%. The new process of preheating shock coupling to improve the forming quality of FDM is proposed, which provides a new method for improving the porosity and mechanical properties of FDM forming samples in the future.

(3) The interlayer properties of formed samples of PEEK/CF composite under thermal coupling were investigated. The results show that the interlayer properties of PEEK/CF composites decrease from 16.9 MPa to 8.9 MPa after thermal coupling because the height of the printing layer is increased after thermal coupling, which reduces the interlayer performance.

Author Contributions: Conceptualization, Q.S. and X.W.; data curation, Q.S. and X.W.; formal analysis, Q.S. and X.W.; funding acquisition, G.Y.; investigation, X.Y.; methodology, Q.S.; project administration, X.W.; interpretation of data for the work, Z.J.; software, G.Y.; writing—original draft, Q.S.; writing—review and editing, X.W. and G.Y. All authors have read and agreed to the published version of the manuscript.

Funding: This research received no external funding.

Institutional Review Board Statement: Not applicable.

Data Availability Statement: Data are contains in the article.

Conflicts of Interest: The authors declare no conflicts of interest.

References

1. Chen, Z.H. *The Investigation of Drilling Performance of Short Fibre-Reinforced PEEK Composites*; Fuzhou University: Fuzhou, China, 2006.
2. Li, G.; Xu, X.H.; Wang, Y.L.; Huang, Y.; Wan, Y.Z. Preparation of short carbon fiber reinforced polyetheretherketone composites and their mechanical properties. In Proceedings of the Foundation, Innovation, and Efficiency: National Composite Materials Academic Conference, Yichang, China, 1 October 2006.
3. Li, N.; Hu, Z.; Huang, Y. Preparation and characterization of nanocomposites of poly(p-phenylene benzobisoxazole) with aminofunctionalized graphene. *Polym. Compos.* **2018**, *39*, 2969–2976. [CrossRef]
4. Rahmanian, S.; Suraya, A.R.; Shazed, M.A.; Zahari, R.; Zainudin, E.S. Mechanical characterization of epoxy composite with multiscale reinforcements: Carbon nanotubes and short carbon fibers. *Mater. Des.* **2014**, *60*, 34–40. [CrossRef]
5. Li, Q.; Zhao, W.; Li, Y.; Yang, W.; Wang, G. Flexural properties and fracture behavior of PEEK/CF in orthogonal building orientation by FDM: Microstructure and mechanism. *Polymers* **2019**, *11*, 656. [CrossRef] [PubMed]
6. Han, X.; Yang, D.; Yang, C.; Spintzyk, S.; Scheideler, L.; Li, P.; Li, D.; Geis-Gerstorfer, J.; Rupp, F. Carbon fiber reinforced PEEK composites based on 3D-printing technology for orthopedic and dental applications. *J. Clin. Med.* **2019**, *8*, 240. [CrossRef] [PubMed]
7. Arif, M.F.; Alhashmi, H.; Varadarajan, K.M.; Koo, J.H.; Hart, A.J.; Kumar, S. Multifunctional performance of carbon nanotubes and graphene nanoplatelets reinforced PEEK composites enabled via FFF additive manufacturing. *Composites* **2020**, *184*, 107625. [CrossRef]
8. Zhao, G.; Qin, M.; Liu, Y.; Qu, X.; Ding, M.; Tian, L.; Liu, Y.; Tian, X.; Liu, Y. Optimizing fused deposition molding process parameters for improving forming strength of polyetheretherketone. *Chin.J. Mech. Eng.* **2020**, *56*, 216–222.
9. Huang, Y.; Tian, X.Y.; Zheng, Z.; Li, D.; Malakhov, A.V.; Polilov, A.N. Multiscale concurrent design and 3D printing of continuous fiber reinforced thermoplastic composites with optimized fiber trajectory and topological structure. *Compos. Struct.* **2022**, *285*, 115241. [CrossRef]
10. Liu, T.; Zhang, M.; Kang, Y.; Tian, X.; Ding, J.; Li, D. Material extrusion 3D printing of polyether ether ketone in vacuum environment: Heat dissipation mechanism and performance. *Addit. Manuf.* **2023**, *62*, 103390. [CrossRef]
11. Tian, X.; Liu, T.; Yang, C.; Wang, Q.; Li, D. Interface and performance of 3D printed continuous carbon fiber reinforced PLA composites. *Compos. Part A Appl. Sci. Manuf.* **2016**, *88*, 198–205. [CrossRef]
12. Tian, X.; Liu, T.; Wang, Q.; Dilmurat, A.; Li, D.; Ziegmann, G. Recycling and remanufacturing of 3D printed continuous carbon fiber reinforced PLA composites. *J. Clean. Prod.* **2016**, *142*, 139. [CrossRef]
13. Luo, M.; Tian, X.; Shang, J.; Zhu, W.; Li, D.; Qin, Y. Impregnation and interlayer bonding behaviors of 3D-printed continuous carbon-fiber-reinforced poly-ether-ether-ketone composites. *Compos. Part A Appl. Sci. Manuf.* **2019**, *121*, 130–138. [CrossRef]
14. Hou, Z.; Tian, X.; Zhang, J.; Zhe, L.; Zheng, Z.; Li, D.; Malakhov, A.V.; Polilov, A.N. Design and 3d printing of continuous fiber reinforced heterogeneous composites. *Compos. Struct.* **2020**, *237*, 111945. [CrossRef]
15. Patricia, P.; Bersee, E.; Beukers, A. Residual stresses in thermoplastic composites-a study of the literature-Part II: Experimental techniques. *Compos. Part A Appl. Sci. Manuf.* **2007**, *38*, 651–665.
16. Dong, L.; Makradi, A.; Ahzi, S.; Remond, Y. Three dimensional tenement finite element analysis of the selective lasers interring process. *J. Mater. Process Technol.* **2009**, *209*, 700–706. [CrossRef]
17. Heat Source Model, Gauss Function Distribution Heat Source Mode. Available online: https://wenku.so.com/d/3d72d2d15ca891d09761c9d9867f0c2e.htm (accessed on 7 December 2021).
18. Sun, Q.; Shan, Z.; Zhan, L.; Wang, S.; Liu, X.; Li, Z.; Wu, S. Warp Deformation Model of Polyetheretherketone Composites Reinforced with Carbon Fibers in Additive Manufacturing. *Mater. Res. Express* **2021**, *8*, 125305. [CrossRef]

Article

Effect of Metastable Intermolecular Composites on the Thermal Decomposition of Glycidyl Azide Polymer Energetic Thermoplastic Elastomer

Chao Sang [1,2,3,*] and Yunjun Luo [4,*]

1 School of Chemistry and Chemical Engineering, Dezhou University, Dezhou 253023, China
2 Shandong Provincial Engineering Research Center of Organic Functional Materials and Green Low-Carbon Technology, Dezhou 253023, China
3 Shandong Provincial Key Laboratory of Monocrystalline Silicon Semiconductor Materials and Technology, College of Chemistry and Chemical Engineering, Dezhou University, Dezhou 253023, China
4 School of Materials Science and Engineering Technology, Beijing Institute of Technology, Beijing 100086, China
* Correspondence: 18810988921@163.com (S.C.); yjluo@bit.edu.cn (L.Y.)

Abstract: Glycidyl azide polymer energetic thermoplastic elastomer (GAP-ETPE) has become a research hotspot due to its excellent comprehensive performance. In this paper, metastable intermolecular energetic nanocomposites (MICs) were prepared by a simple and safe method, and the catalytic performance for decomposition of GAP-ETPE was studied. An X-ray diffraction (XRD) analysis showed that the MICs exhibited specific crystal formation, which proved that the MICs were successfully prepared. Morphology, surface area, and pore structure analysis showed that the Al/copper ferrite and Al/Fe$_2$O$_3$ MICs had a large specific surface area mesoporous structure. The Al/CuO MICs did not have a mesoporous structure or a large surface area. The structure of MICs led to their different performance for the GAP-ETPE decomposition catalysis. The increase in specific surface area is a benefit of the catalytic performance. Due to the easier formation of complexes, MICs containing Cu have better catalytic performance for GAP-ETPE decomposition than those containing Fe. The conclusions of this study can provide a basis for the adjustment of the catalytic performance of MICs in GAP-ETPE propellants.

Keywords: metastable intermolecular composite; thermal decomposition; GAP-ETPE

1. Introduction

Energetic thermoplastic elastomers (ETPEs) are used as solid propellant binders to impart advantages such as high energy, blunt sensitivity, low characteristic signal, and recyclability of propellants [1]. Energetic thermoplastic elastomers refer to thermoplastic elastomers containing energetic groups such as nitrate group (-ONO$_2$), nitro group (-NO$_2$), nitroamine group (-NNO$_2$), azide group (-N$_3$), and difluoroamine group (-NF$_2$) [2]. Among them, azide energetic thermoplastic elastomers have attracted extensive attention due to their high heat release, no need for oxygen consumption during decomposition, and good compatibility with nitroamine explosives, among which polyazide glycidyl ether (GAP)-based ETPE is the representative. GAP-based ETPE (GAP-ETPE) has a soft-hard segment structure, in which the soft segment is partially or wholly composed of GAP, which gives ETPE high energy and thus increases the combustion rate of GAP-ETPE propellant, and the hard segment is composed of urethane segments, which gives ETPE high tensile strength (tensile strength is about 4.8 MPa, elongation at break is about 580% [3]), so GAP-ETPE propellants have become a research hotspot in thermoplastic elastomer propellants [4]. As the propellant binder, the thermal decomposition performance of GAP-ETPE will seriously affect the combustion performance of the propellant and then affect the thrust of the propellant to the rocket, so it is essential to catalyze its thermal decomposition. Research shows that the decomposition and combustion of GAP-ETPE can be catalyzed by transition metal oxides. And the catalysis to the binder (GAP-ETPE) can significantly improve the combustion performance of the propellant. Li found that CuO and $Cr_2Cu_2O_5$ can drastically reduce the decomposition temperature of the azide group [5].

Metastable intermolecular composites (MICs) have short fuel-oxidizer diffusion distance; hence, they have a shorter ignition time, a higher burning speed, and a faster reaction speed than traditional thermites [6–8]. MICs are composed of oxidizer particles (such as Fe_2O_3, CuO, Bi_2O_3, etc.) and nano-sized metal fuel (mostly Al) [9–12], which are also termed nano-thermites, nanocomposite energetic materials, super thermites, metastable intermixed composites, etc. MICs are widely used in airbag ignition materials, propellants, aerospace equipment, pyrotechnics, pressure-mediated molecular transfer, and biomedical-related applications, both in civilian and military applications [13–15]. In addition to the excellent reactivity, MICs can also play the role of combustion catalyst when applied in solid propellants. The transition metal oxides such as CuO, Fe_2O_3 and Bi_2O_3 in the above-mentioned MICs can give MICs catalytic properties to solid propellant combustion [16–19]. At present, there are no studies on the effect of the transition metal oxides in MICs on the decomposition of GAP-ETPE.

In this work, Al/copper ferrites (include $CuFe_2O_4$, CuO and Fe_2O_3) MICs were prepared by a facile sol-gel reaction under mild conditions (low temperature calcination). We studied the morphology and structure changes of copper ferrites and Al/copper ferrites MICs, and the influence of these changes on their catalytic performance for GAP-ETPEE decomposition.

2. Experiment

2.1. Chemicals

Hexamethylene di-isocyanate (HMDI) from Bayer. Co. (Leverkusen, Germany). was used without any treatment. A glycidyl azide polymer (GAP; OH equivalent: 26.71 mg KOH/g) from Liming Research Institute of Chemical Industry (Luoyang, China) was used after vacuum drying for 2 h at 90 °C. A polymerization catalyst (dibutyltin dilaurate, DBTDL) from Beijing Chemical Plant (Beijing, China) was dissolved into dibutyl phthalate after using. Also, 1,4-butanediol (BDO) from Beijing Chemical Plant was vacuum dried for 4 h at 85 °C. One of the bonding agents, N-(2-cyanoethyl) diethanolamine (CBA), was made in our laboratory [20]. Acrylonitrile (Beijing Tong Guang Fine Chemicals Company, Beijing, China) was purified by vacuum distillation. Copper nitrate trihydrate ($Cu(NO_3)_2 \cdot 3H_2O$), diethanolamine, iron nitrate nonahydrate ($Fe(NO_3)_3 \cdot 9H_2O$), 1,2-propylene oxide, absolute ethanol, n-hexane, ethyl acetate, N,N-Dimethylformamide, which are all analytical grade, were purchased from Beijing Tong Guang Fine Chemicals Company (Beijing, China) and used without further purification.

2.2. Preparation of Samples

The whole preparation route of the MICs/GAP-ETPE samples is illustrated in Figure 1.

(1) Preparation of GAP-ETPE

The preparation of the glycidyl azide polymer energetic thermoplastic elastomer (GAP-ETPE) was achieved by the method reported in the literature [3]. The chain extenders were CBA (whose -CN group gives GAP-ETPE the molecular bonding function) and BDO. The number-average molecular weight (Mn) of GAP-ETPE was about 29,800 $g \cdot mol^{-1}$, and the ratio of hard segment and soft segment was 3:7.

(2) Preparation of MICs

The Al/copper ferrite MICs was prepared via the method in the reference [21]. The scale of n(Cu) and n(Fe) was 1:4, and n(Al)/(n(Fe) + n(Cu)) was 3, which were proven to be the best ratio in our previous research. The products were labeled as CFA. As a comparison, Al/Fe_2O_3 and Al/CuO MICs were prepared by the same method. The same amount of $Cu(NO_3)_2 \cdot 3H_2O$ (or $Fe(NO_3)_3 \cdot 9H_2O$) were replaced by $Fe(NO_3)_3 \cdot 9H_2O$ (or

$Cu(NO_3)_2 \cdot 3H_2O$. The products were labeled as FA and CA. Furthermore, without the position of Al, copper ferrite, Fe_2O_3, and CuO were prepared by the same method, which were labeled as CF, F, and C. The label and preparation information of the products are shown in Table 1.

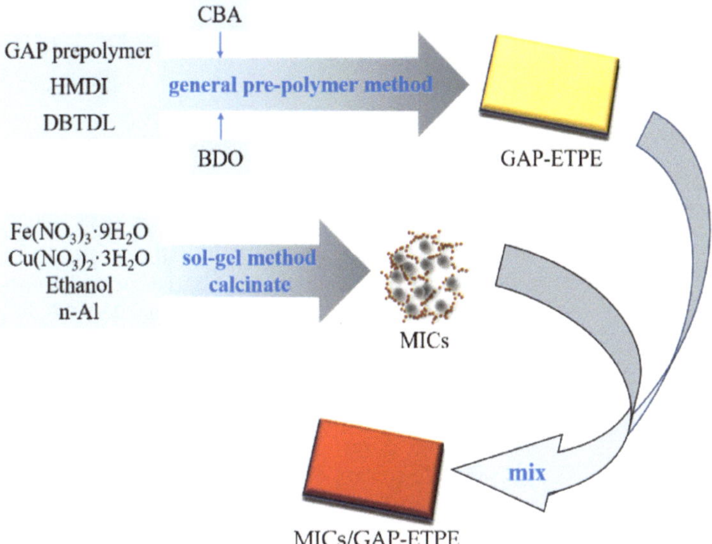

Figure 1. Schematic preparation process of MICs/GAP-ETPE.

Table 1. The label and preparation information of the products.

Label	Product	Raw Material (mol)		
		$Fe(NO_3)_3 \cdot 9H_2O$	$Cu(NO_3)_2 \cdot 3H_2O$	Al
F	Fe_2O_3	0.060	0	0
C	CuO	0	0.060	0
CF	copper ferrite	0.012	0.048	0
FA	Al/Fe_2O_3 MICs	0.060	0	0.180
CA	Al/CuO MICs	0	0.060	0.180
CFA	Al/copper ferrite MICs	0.012	0.048	0.180

(3)   Preparation of MICs/GAP-ETPE

First, the GAP-ETPE was dissolved in tetrahydrofuran. Afterwards, the MICs powder was added to the solution at the percentage of 1% and sonicated for 30 min. The solvent was then evaporated in a vacuum oven at 60 °C for 6 h. Finally, the mixture was mixed using an open mill. The products prepared with F, C, CF, FA, CA, and CFA were, respectively, denoted as GF, GC, GCF, GFA, GCA, and GCFA. In particular, the sample without a catalyst was named G0.

2.3. Measurements and Characterizations

A S4800 cold field scanning electron microscope (SEM) (Japan's Hitachi Corporation, Tokyo, Japan) was used to observe the morphology of the MICs at an accelerating voltage of 15.0 kV. An ASAP 2020 volumetric analyzer (Micromeritics Instrument Corporation, Norcross, GA, USA) was employed to investigate the surface area and pore structure of the MICs (degassed at 120 °C for at least 6 h). An X'Pert Pro MPD (PANalytical, Netherlands) diffractometer with monochromatic Cu Ka radiation (λ = 1.5406 Å) at 40 kV and 40 mA was used to carry out the X-ray diffraction (XRD) measurements of the MICs with a scanning speed of $0.01° s^{-1}$ and a step size of 0.01 from $10°$ to $90°$ (2θ). A METTLER TOLEDO TGA/DSC 1-Thermogravimetric Analyzer (Zurich, Switzerland) was used to measure the thermal performance of GAP-ETPE under the conditions of an ultrapure nitrogen atmosphere, an uncovered alumina ceramic crucible, a heating rate of $10 °C min^{-1}$, and a heating range of $30 °C$ to $600 °C$.

3. Results and Discussion

3.1. Characters of MICs

3.1.1. Crystal Forms

Phase investigations of the crystallized products were performed by XRD and the powder diffraction patterns were presented in Figure 2. As shown in Figure 2a, the diffraction pattern of C, F, and CF can be indexed to CuO (standard PDF card NO. 48-1548), γ-Fe_2O_3 crystal (standard PDF card NO. 25-1402) and $CuFe_2O_4$ (standard PDF card NO. 25-0283) [22]. Crystal defects in copper ferrite crystals are more than those in the pure iron oxide and copper oxide crystals. Crystal defects are beneficial to its catalytic function [23].

It can also be found from Figure 2 that both CA, FA, and CFA have the same set of additional diffraction peaks, indicating that the existence of Al (standard PDF card NO. 04-0787) in the products and demonstrate the sol-gel process could hold the structure of Al nanocrystalline particles. The diffraction peaks intensity of Al in CA are higher than those in FA and CFA, which indicates more Al is exposed on the outside of the oxide structure, which can also be proven in the following SEM images. The relative content of CuO, γ-Fe_2O_3, or $CuFe_2O_4$ reduces with the existence of Al, so their diffraction peak gradually weakens. The existence of n-Al does not change the crystal form of CuO, γ-Fe_2O_3, or $CuFe_2O_4$. This proves that the MICs were successfully prepared.

3.1.2. Morphological Characterization

The surface morphologies of the MICs were observed by SEM, as shown in Figure 3. It can be seen from the figures that F and CF have a rough surface, while the surface morphologies of C are relatively smooth. The sample surface became rougher as the Al was added. The particles with a diameter of about 90 nm, that is, n-Al particles, can be clearly observed in the structure of the CA. But the same particles were not observed in the FA and CA, which indicates that the n-Al particles exist inside the structure of Fe_2O_3 and $CuFe_2O_4$. And this composite form that exposes oxides to the outermost layer is conducive to increasing the contact probability between the catalyst and the catalyzed substance and improving the catalytic activity when used as a catalyst for other reactions.

3.1.3. Specific Surface Area and Pore Volume

As shown in Figure 4, the nitrogen adsorption-desorption isotherms of both Fe_2O_3, copper ferrites, Al/copper ferrite MICs, and Al/Fe_2O_3 MICs belong to type IV, which means abundant mesopores in the structure of the samples. But the CuO and Al/CuO MICs show little mesopores. The measured specific surface areas (S_{BET}) of samples are presented in Table 2. It can be seen from the table that the specific surface area of Al/copper ferrite MICs and Al/Fe_2O_3 MICs are similar and show similar specific surface area changes with the presence of n-Al. The specific surface area of the MICs rise sharply after the addition of n-Al. It is because, as part of the gel skeleton, the n-Al increased the porosity of the MICs.

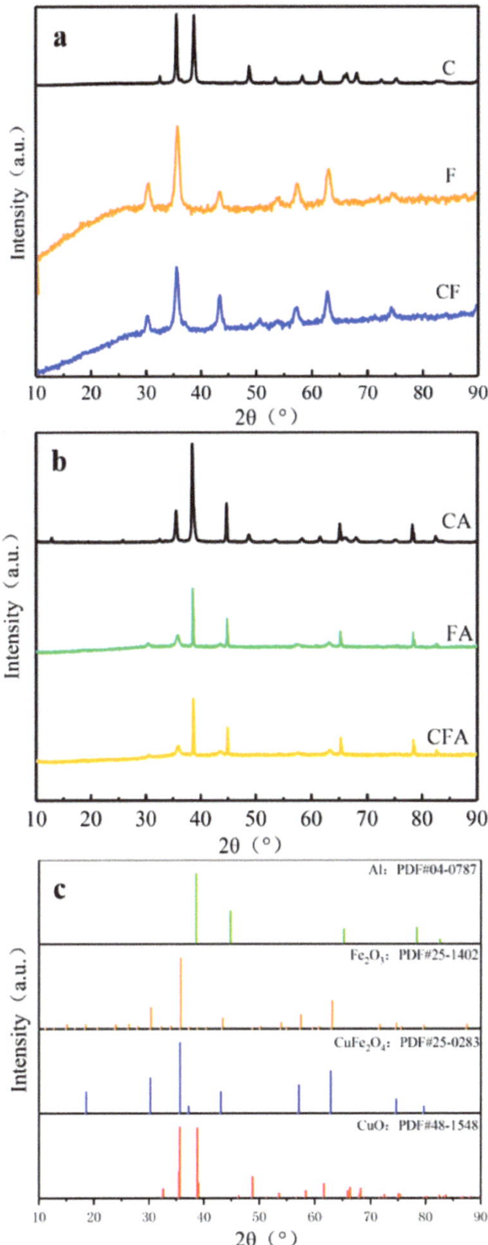

Figure 2. X-ray diffraction patterns of copper ferrites: (**a**) MICs (**b**) and PDF cards of Al, Fe_2O_3, $CuFe_2O_4$, CuO (**c**).

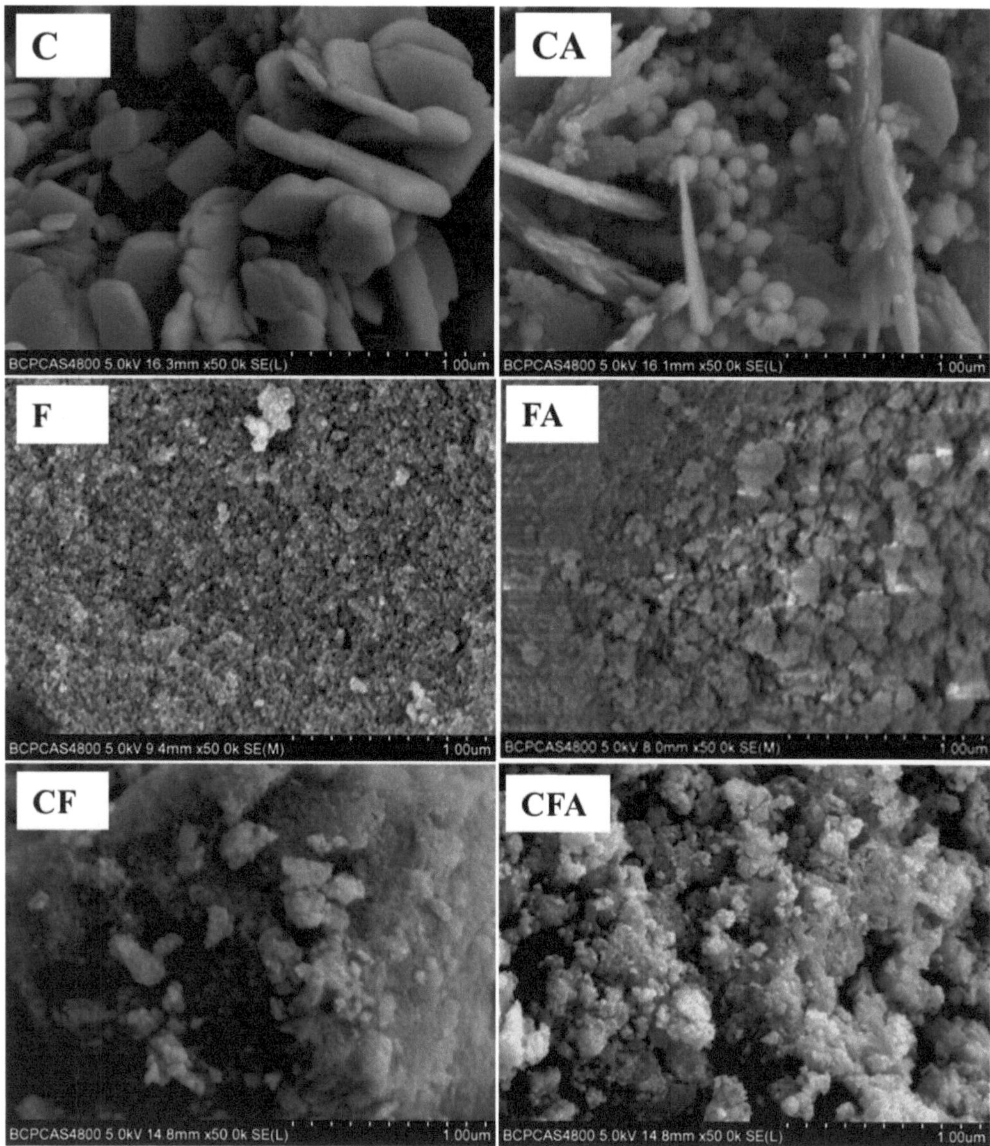

Figure 3. SEM images of Al/copper ferrite MICs.

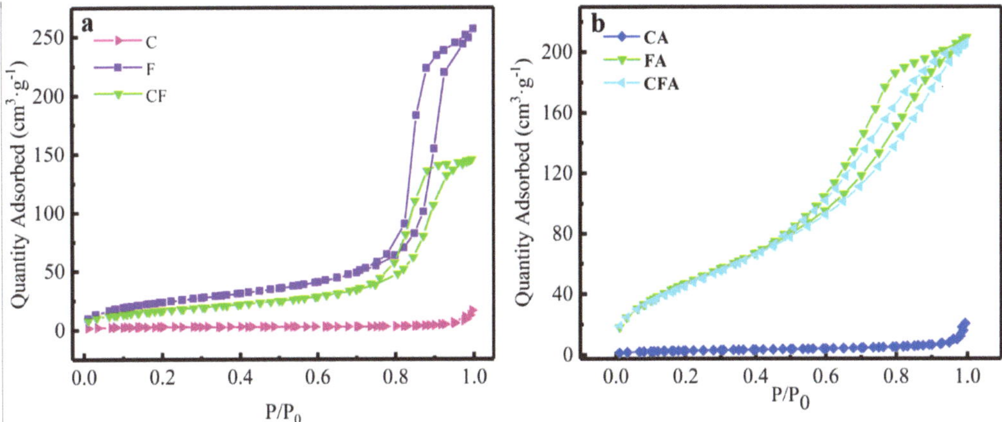

Figure 4. Nitrogen adsorption-desorption isotherms of copper ferrites (**a**) and MICs (**b**).

Table 2. Nitrogen adsorption-desorption isotherms parameters of MICs.

Samples	S_{BET}/m^2 g^{-1}	V_{tot}/cm^3 g^{-1}	D_{ave}/nm
C	2.88	0.02	29.63
CA	6.39	0.03	15.92
F	87.92	0.39	17.07
FA	178.21	0.32	6.26
CF	80.43	0.26	11.77
CFA	170.55	0.31	6.39

Note: S_{BET} is the specific surface area calculated by the BET method, V_{tot} is the total pore volume, and D_{ave} is the average pore diameter.

Generally, materials which have large specific surface areas have an excellent catalytic performance. That is, they benefit from the large number of catalytic active sites. With the same n-Al content, the pore volume and specific surface area of the Al/copper ferrite MICs are comparable to those of the Al/Fe$_2$O$_3$ MICs.

3.2. Thermal Decomposition of GAP-ETPE

GAP-ETPE is a promising solid propellant binder and its thermal decomposition performance can significantly affect the combustion performance of propellants. The transition metal oxide contained in MICs can be used to catalyze the thermal decomposition of GAP-ETPE. With the help of the TG and DSC technique, we discussed the effects of MICs for the thermal decomposition of GAP-ETPE.

Figure 5a,b shows the TG and DTG curves of the MICs/GAP-ETPE, and Table 3 lists the relevant parameters. Figure 5c shows the thermal effects during the MICs/GAP-ETPE thermal decomposition, which are tested by DSC. In order to facilitate comparison, the data of pure GAP-ETPE are also presented in the above figures and tables. All of the samples exhibit three weight losses, corresponding to the three decomposition stages of GAP-ETPE [24]. The first weight loss stage is at 200~90 °C, and the weight loss mass fraction was about 30%, which was basically consistent with the mass fraction of the azide group (29.7%) in the GAP-based ETPE. At the same time, the first stage decomposition peak temperature (T_{P1}) in DTG corresponds to the exothermic peak temperature (T_P) in the DSC curves. Therefore, the first weight loss stage corresponds to the side-chain azide. The second weightless stage appears at 280~380 °C, which corresponds to the urethane segment formed by HMDI, CBA and BDO. The third thermal decomposition stage shows at 380~490 °C, which corresponds to the decomposition of the polyether backbone, is related to the literature report on the thermal decomposition of GAP-based ETPEs [20].

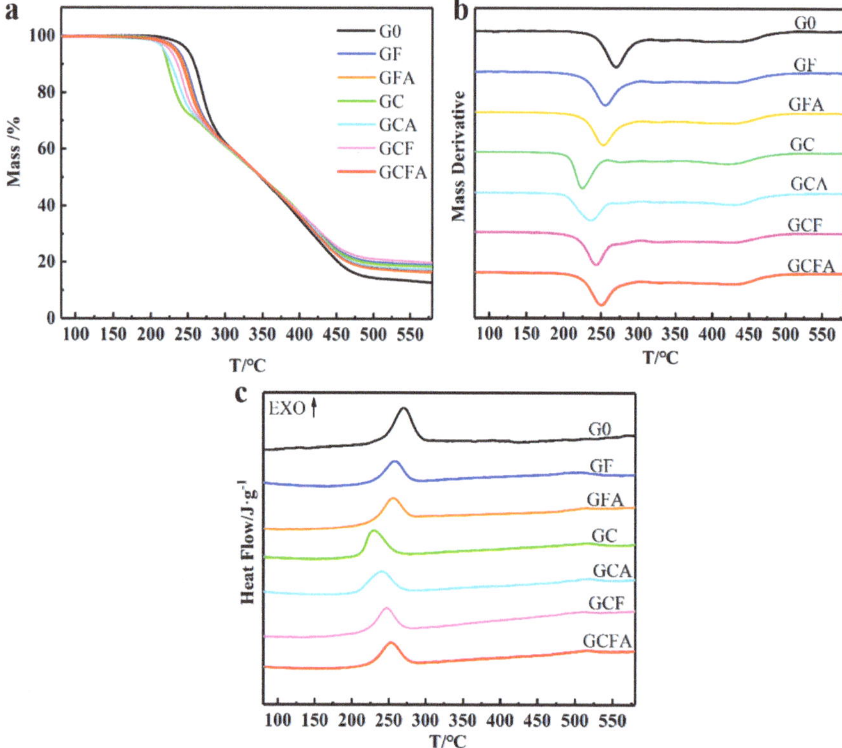

Figure 5. TG (**a**), DTG (**b**), and DSC (**c**) curves of MICs/GAP-ETPE.

Table 3. TG and DSC parameters of MICs/GAP-ETPE.

Samples	T_{P1} (°C)	T_P (°C)
G0	268.9	270.4
GC	223.2	227.5
GCA	235.5	239.9
GF	255.1	256.6
GFA	252.2	254.4
GCF	243.5	246.4
GCFA	250.1	252.9

The thermal decomposition process of GAP-ETPE is shown in Figure 6. The first step of the thermal decomposition of azide groups is the breaking of the RN-N$_2$ bond to generate nitrene, followed by the rearrangement of nitrogen to generate imine, and the release of N$_2$ imine to generate NH$_3$ through a H transfer and free radical transfer, or the breaking of the C-C bond to generate HCN, etc. The formation of NH$_3$ is an exothermic reaction, and the formation of HCN is an endothermic reaction.

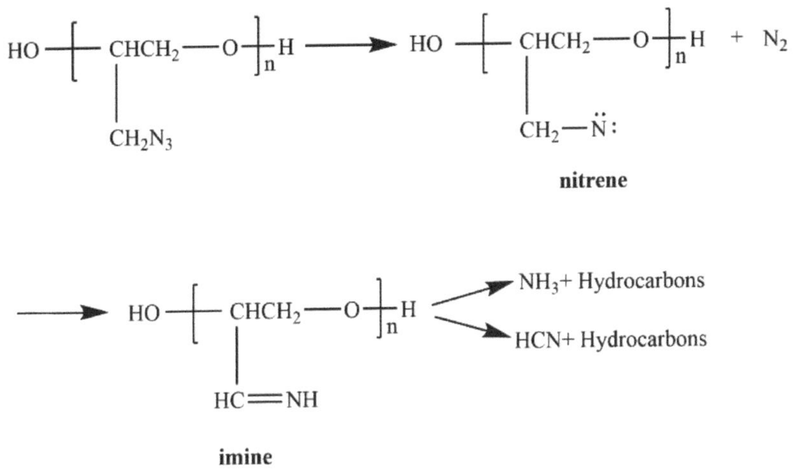

$$\text{HO}\!-\!\!\left[\text{CHCH}_2\!-\!\text{O}\right]_n\!\!-\!\text{H} \quad\longrightarrow\quad \text{HO}\!-\!\!\left[\text{CHCH}_2\!-\!\text{O}\right]_n\!\!-\!\text{H} \;+\; \text{N}_2$$

$$\underset{\text{CH}_2\text{N}_3}{\big|} \qquad\qquad\qquad\qquad \underset{\text{CH}_2\!-\!\ddot{\text{N}}:}{\big|}$$

nitrene

$$\longrightarrow\quad \text{HO}\!-\!\!\left[\text{CHCH}_2\!-\!\text{O}\right]_n\!\!-\!\text{H} \;\underset{\searrow}{\overset{\nearrow}{}}\; \begin{array}{l}\text{NH}_3\text{+ Hydrocarbons}\\[4pt]\text{HCN+ Hydrocarbons}\end{array}$$

$$\underset{\text{HC}=\!\!=\text{NH}}{\big|}$$

imine

Figure 6. Thermal decomposition process of GAP-ETPE.

It can be seen from Figure 5 that the thermogravimetric curves of the different samples are similar. The thermal decomposition mechanism of the GAP-ETPE has not been changed by the addition of MICs, but each decomposition steps temperature was changed. The azide maximum decomposition rate temperature (T_{P1}) of pure GA-ETPE is 268.9 °C. The T_{P1} declines obviously after the addition of MICs, indicating that the MICs significantly affect azide decomposition.

As shown in Figure 5, the T_{P1} of GC and GCA are lower, and those of GF and GFA are higher. The reason is that the activation center of CuO can form an activation complex with the azide group, which reduces the activation energy of the decomposition and promotes the cleavage of RN-N$_2$ bonds to form nitrogen bins, and N$_2$ is also released, which advances the peak temperature of the decomposition of the azide group. The ability of Fe$_2$O$_3$ to form an activation complex is lower than that of CuO. Due to the presence of the Cu element, the catalytic capacity of CF and CFA for the decomposition of azides is stronger than that of F and FA.

The T_{P1} of GFA is lower than that of GF, which is because the more catalytic active sites provided by the huge specific surface area of FA. However, GC and GCA show the opposite pattern. The reasons are obviously as follows: The presence of n-Al did not change the specific surface area of the Al/CuO MICs so that there are no more catalytic active sites in Al/CuO MICs. Furthermore, the self-heating thermal decomposition process of the GAP-ETPE will be slowed down by the "dilution" and heat-conduction effects of aluminum [25]. Because of the dilution to the Cu element, the catalytic ability of CFA is lower than CF. But thanks to a large increase in specific surface area (80.43 to 170.55 m^2 g^{-1}, see Table 2), the T_{P1} of GCFA is only 6.6 °C higher than that of GCF. In conclusion, the catalytic ability of the catalyst can be improved by controlling the morphology. Besides, there seems to be no obvious synergistic catalytic effect of Cu-Fe oxides for the decomposition of GAP-ETPE.

4. Conclusions

Novel MICs successfully prepared by simple and mild methods were employed for the thermal decomposition of GAP-ETPE. The prepared Al/copper ferrite and Al/Fe$_2$O$_3$ metastable intermolecular energetic nanocomposites had a mesoporous structure with a large specific surface area, and the specific surface area greatly increased with the addition of n-Al. The Al/CuO MICs did not have a mesoporous structure and the specific surface area is very small. The structure of MICs led to their different performance for the GAP-ETPE decomposition catalysis. The increase in the specific surface area can improve the catalytic performance of the GAP-ETPE decomposition. Due to the easier complex formation, Cu-containing MICs have a better catalytic performance for GAP-ETPE decomposition than Fe-containing MICs. The MICs can reduce the decomposition temperature of the GAP-ETPE azide group by up to 45.7 °C. The conclusion of this study can help promote the application of GAP-ETPE as the binder in solid propellant.

Author Contributions: C.S.—conceptualization, formal analysis, funding acquisition, investigation, methodology, project administration, resources, validation, writing; Y.L.—conceptualization, supervision. All authors have read and agreed to the published version of the manuscript.

Funding: This research was funded by the Nature Science Foundation of Shandong Province (ZR2023QB294, ZR2023QB190) and the Doctoral Foundation of Dezhou University (2022xjrc305, 2022xjrc105).

Institutional Review Board Statement: Not applicable.

Data Availability Statement: Data are contained within the article.

Conflicts of Interest: The authors declare no conflict of interest.

References

1. Talawar, M.B.; Sivabalan, R.; Mukundan, T.; Muthurajan, H.; Sikder, A.K.; Gandhe, B.R.; Subhananda Rao, A. Environmentally compatible next generation green energetic materials (GEMs). *J. Hazard. Mater.* **2009**, *161*, 589–607. [CrossRef] [PubMed]

2. Arun Sikder, K.; Reddy, S. Review on energetic thermoplastic elastomers (ETPEs) for military science. *Propellants Explos. Pyrotech.* **2013**, *38*, 14–28. [CrossRef]

3. Zhang, Z.; Wang, G.; Wang, Z.; Zhang, Y.; Ge, Z.; Luo, Y. Synthesis and characterization of novel energetic thermoplastic elastomers based on glycidyl azide polymer (GAP) with bonding functions. *Polym. Bull.* **2015**, *72*, 1835–1847. [CrossRef]

4. Badgujar, D.M.; Talawar, M.B.; Asthana, S.N.; Mahulikar, P.P. Advances in science and technology of modern energetic materials: An overview. *J. Hazard. Mater.* **2008**, *151*, 289–305. [CrossRef] [PubMed]

5. Li, X.; Ge, Z.; Li, Q.; Li, D.; Zuo, Y.; Yan, B.; Luo, Y. Effect of Burning Rate Catalysts on the Thermal Decomposition Properties of GAP-based ETPE Energetic Thermoplastic Elastormer. *Chin. J. Energetic Mater.* **2016**, *24*, 1102–1107. [CrossRef]

6. Farley, C.; Pantoya, M. Reaction kinetics of nanometric aluminum and iodine pentoxide. *J. Therm. Anal. Calorim.* **2010**, *102*, 609–613. [CrossRef]

7. Pantoya, M.L.; Granier, J.J. The effect of slow heating rates on the reaction mechanisms of nano and micron composite thermite reactions. *J. Therm. Anal. Calorim.* **2006**, *85*, 37–43. [CrossRef]

8. Sun, J.; Pantoya, M.L.; Simon, S.L. Dependence of size and size distribution on reactivity of aluminum nanoparticles in reactions with oxygen and MoO_3. *Thermochim. Acta.* **2006**, *444*, 117–127. [CrossRef]

9. Elbasuney, S.; Hamed, A.; Yehia, M.; Ramzy, H.; Mokhtar, M. The Impact of Metastable Intermolrecular Nanocomposite Particles on Kinetic Decomposition of Heterocyclic Nitramines Using Advanced Solid-Phase Decomposition Models. *J. Inorg. Organomet. Polym. Mater.* **2021**, *31*, 3665–3676. [CrossRef]

10. He, W.; Ao, W.; Yang, G.; Yang, Z.; Guo, Z.; Liu, P.; Yan, Q. Metastable energetic nanocomposites of MOF-activated aluminum featured with multi-level energy releases. *Chem. Eng. J.* **2020**, *381*, 122623. [CrossRef]

11. Song, Z.; Jin, M.; Xian, M.; Wang, C. Peptide-driven assembly of Al/CuO energetic nanocomposite material. *Chem. Eng. J.* **2020**, *338*, 124225. [CrossRef]

12. Elbasuney, S.; Hamed, A.; Ismael, S.; Mokhtar, M.; Gobara, M. Novel high energy density material based on metastable intermolecular nanocomposite. *J. Inorg. Organomet. Polym. Mater.* **2020**, *30*, 3980–3988. [CrossRef]

13. Parimi, V.S.; Huang, S.; Zheng, X. Enhancing ignition and combustion of micronsized aluminum by adding porous silicon. *Proc. Combust. Inst.* **2017**, *36*, 2317–2324. [CrossRef]

14. Thiruvengadathan, R.; Staley, C.; Geeson, J.M.; Chung, S.; Raymond, K.E.; Gangopadhyay, K.; Gangopadhyay, S. Enhanced combustion characteristics of bismuth trioxide-aluminum nanocomposites prepared through graphene oxide directed self-assembly. *Propellants Explos. Pyrotech.* **2015**, *40*, 729–734. [CrossRef]

15. Kim, S.B.; Kim, K.J.; Cho, M.H.; Kim, J.H.; Kim, K.T.; Kim, S.H. Micro- and nanoscale energetic materials as effective heat energy sources for enhanced gas generators. *ACS Appl. Mater. Interfaces* **2016**, *8*, 9405–9412. [CrossRef]

16. Fehlberg, S.; Örnek, M.; Manship, T.D.; Son, S.F. Decomposition of ammonium-perchlorate-encapsulated nanoscale and micronscale catalyst particles. *J. Propuls. Power* **2020**, *36*, 862–868. [CrossRef]

17. Chaturvedi, S.; Dave, P.N. Nano-metal oxide: Potential catalyst on thermal decomposition of ammonium perchlorate. *J. Exp. Nanosci.* **2012**, *7*, 205–231. [CrossRef]

18. Reese, D.A.; Son, S.F.; Groven, L.J. Preparation and characterization of energetic crystals with nanoparticle inclusions. *Propellants Explos. Pyrotech.* **2012**, *37*, 635–638. [CrossRef]

19. Alizadeh-Gheshlaghi, E.; Shaabani, B.; Khodayari, A. Investigation of the catalytic activity of nano-sized CuO, Co_3O_4 and $CuCo_2O_4$ powders on thermal decomposition of ammonium perchlorate. *Powder Technol.* **2012**, *217*, 330–339. [CrossRef]

20. Chen, L.; Deng, J.; Xu, W.; Zhao, D.; Yang, W.; Yu, H. Synthesis of N-(2-Cyanoethyl) Diethanolamine. *Petrochem. Technol.* **2007**, *36*, 1029–1031.
21. Sang, C.; Chen, K.; Li, G.; Jin, S.; Luo, Y. Facile mass preparation and characterization of Al/copper ferrites metastable intermolecular energetic nanocomposites. *RSC Adv.* **2021**, *11*, 7633–7643. [CrossRef] [PubMed]
22. Tasca, J.E.; Quincoces, C.E.; Lavat, A. Preparation and characterization of $CuFe_2O_4$ bulk catalysts. *Ceram. Int.* **2011**, *37*, 803–812. [CrossRef]
23. Yan, Q.L.; Zhao, F.Q.; Kuo, K.K.; Zhang, X.H.; Zeman, S.; Deluca, L.T. Catalytic effects of nano additives on decomposition and combustion of RDX-, HMX-, and AP-based energetic compositions. *Prog. Energy Combust. Sci.* **2016**, *57*, 75–136. [CrossRef]
24. Boldyrev, V.V.; Alexandrov, V.V.; Boldyreva, A.V.; Gritsan, V.I.; Karpenko, Y.Y.; Korobeinitchev, O.P.; Panfilov, V.N.; Khairetdinov, E.F. On the mechanism of the thermal decomposition of ammonium perchlorate. *Combust. Flame* **1970**, *15*, 71–77. [CrossRef]
25. Liu, Z.R. *Thermal Analyses for Energetic Materials*; National Defense Industry Press: Beijing, China, 2008.

Article

Mechanical, Dielectric and Flame-Retardant Properties of GF/PP Modified with Different Flame Retardants

Jingwen Li [1], Yiliang Sun [1,*], Boming Zhang [1] and Guocheng Qi [2]

1   School of Materials Science and Engineering, Beihang University, Beijing 100191, China; ljw666@buaa.edu.cn (J.L.); zbm@buaa.edu.cn (B.Z.)
2   Department of Mechanics, Beijing Jiaotong University, Beijing 100044, China; gchqi@bjtu.edu.cn
*   Correspondence: ylsun@buaa.edu.cn

Abstract: With the rapid development of electronic information technology, higher requirements have been put forward for the dielectric properties and load-bearing capacity of materials. In continuous glass fiber-reinforced thermoplastic composites, polypropylene matrix is a non-polar polymer with a very low dielectric constant and dielectric loss, but polypropylene is extremely flammable which greatly limits its application. Aiming at the better application of flame retardant-modified continuous glass fiber-reinforced polypropylene composites (FR/GF/PP) in the field of electronic communication, the effects of four different kinds of flame retardants (Decabromodiphenyl ethane (DBDPE), halogen-free one-component flame retardant (MONO), halogen-free compound flame retardant (MULTI), and intumescent flame retardant (IFR)) on the properties of FR/GF/PP were compared, including the mechanical properties, dielectric properties and flame-retardant properties. The results showed that among the FR/GF/PP, IFR has the highest performance in mechanical properties, MULTI has better performance in LOI, DBDPE and IFR have better performance in flame retardant rating, and DBDPE and IFR have lower dielectric properties. Finally, gray relational analysis is applied to propose an approach for selecting the optimal combination (flame retardant type and flame-retardant content) of comprehensive performance. In the application exemplified in this paper, the performance of IFR-3-modified GF/PP is optimized.

Keywords: flame retardant-modified glass fiber-reinforced polypropylene; mechanical properties; dielectric properties; flame-retardant properties; gray relational analysis

Citation: Li, J.; Sun, Y.; Zhang, B.; Qi, G. Mechanical, Dielectric and Flame-Retardant Properties of GF/PP Modified with Different Flame Retardants. *Polymers* **2024**, *16*, 1681. https://doi.org/10.3390/polym16121681

Academic Editor: Bob Howell

Received: 29 May 2024
Revised: 11 June 2024
Accepted: 12 June 2024
Published: 13 June 2024

1. Introduction

With the rapid development of electronic information technology, the technical strength and accuracy of antenna systems [1,2] are becoming higher, the frequency of electromagnetic wave transmission and reception is becoming higher, and the signal transmission is becoming faster [3–5]. This imposes higher requirements on the wave-transparent properties of materials. For wave-transparent materials, the larger the dielectric constant or the thicker the material, the greater the reflection of electromagnetic waves, and the signal transmission efficiency will be reduced [6–8]. The larger the dielectric loss tangent tan δ is, the more energy is lost, as the electromagnetic wave energy is converted to heat in the process of transmitting through the material [9,10]. Therefore, it is required that the thickness of the wave-transmitting material is as small as possible, the dielectric constant is as low as possible, and the loss angle tangent is as low as close to zero in order to minimize reflection and maximize transmission. This places high demands on the dielectric and mechanical properties of the materials [11].

Thermoplastic composites, as a major branch of composites, have significant advantages such as rapid prototyping, being repairable, having secondary molding, and being recyclable and environmentally friendly, in addition to the advantages of light weight and high strength [12–14]. Continuous glass fiber (GF) reinforced thermoplastic composites with excellent performance and a low price are widely used in aerospace, building materials, vehicles and other fields. The polypropylene (PP) matrix is a non-polar polymer with a very low dielectric constant and dielectric loss, and its dielectric properties remain stable under variable temperature and frequency, while polypropylene has low water absorption (water absorption increases the dielectric properties) [15–18]. Polypropylene's excellent wave-transparent properties make it ideally suited in wave-transparent materials. However, PP has poor flame retardancy, which greatly limits its application [19–21]. Continuous fiber reinforced polypropylene composites (GF/PP) with high mechanical properties, good flame-retardant properties, low dielectric constant and low dielectric loss are beneficial to its application in wave-transparent materials. At present, the research on the properties of FR/GF/PP mostly stays in the study of flame-retardant properties and mechanical properties [22–25], and there is less research on dielectric properties. In order to utilize the FR/GF/PP for better applications in the field of electronic communication, such as in radomes, cabinet casings, plastic vibrators, reflector plates, filters, etc., more studies are needed to be carried out on the dielectric properties of the FR/GF/PP. In previous studies [26], it was found that a large amount of flame-retardant addition would lead to an increase in the dielectric constant and dielectric loss of GF/PP, but there are many types of flame retardants and the effects of different types of flame retardants on the performance of FR/GF/PP may be different, so the question of how to choose the flame retardants needs to be solved urgently.

In this paper, four types of flame retardants which are relatively common in engineering applications were selected. The effects of the types and contents of these four flame retardants on the mechanical, dielectric and flame-retardant properties of FR/GF/PP were compared. Finally, the gray relational analysis [27–29] was applied to transform the multi-objective, multi-response problem into a single gray relational comparison problem in order to obtain the flame retardant type and the content that would make the comprehensive performance of FR/GF/PP optimal. It provides a reference for the practical application of FR/GF/PP.

2. Materials and Methods

2.1. Materials

Polypropylene (PP, BX3920) with a melt flow index of 100 g/10 min (2.16 kg at 230 °C) was supplied by SK Group (Seoul, Republic of Korea). High-melt-index polypropylene (HPP, MF650X) with a melt flow index of 1200 g/10 min (2.16 kg at 230 °C) was supplied by LyondellBasell Corporation (Jubail, Saudi Arabia). Maleic anhydride-grafted polypropylene (MAPP, Orevac® CA100) was supplied by Arkema (Paris, France). Four commercially available flame retardants (Decabromodiphenyl ethane (DBDPE), halogen-free one-component flame retardant (MONO), halogen-free compound flame retardant (MULTI), and intumescent flame retardant (IFR)), which are commonly used in polypropylene, were supplied by Yantai Xinxiu Corporation (Yantai, China). Direct roving of glass fiber (GF, 4305S) was supplied by the Chongqing Polycomp International Corporation (Chongqing, China).

2.2. Preparation of FR/PP

PP, PP-MF650X, MAPP, and FR were poured into the mixer according to the ratios in Table 1, and mixed well. The homogeneous mixture was poured into the twin-screw extruder for pelletizing, in order to mix the flame retardant homogeneously, and the pelletizing was repeated 3 times.

Table 1. Designation and composition of FR/PP.

Samples	PP (phr)	MAPP (phr)	HPP (phr)	FR (phr)	FR Content (wt%)
0	100	5	10	0	0
X-1	100	5	10	10	8.00
X-2	100	5	10	20	14.81
X-3	100	5	10	30	20.69

Where X is flame retardant (DBDPE, MONO, MULTI, IFR) and flame-retardant content = flame retardant/ (PP + MAPP + HPP + flame retardant).

2.3. Preparation of FR/GF/PP Laminates

A schematic diagram of the preparation process of FR/GF/PP composite laminates is shown in Figure 1. FR/GF/PP prepregs were prepared using a continuous fiber-reinforced thermoplastic composite prepreg production line homemade in the laboratory. FR/PP pellets were melted at the extruder head position and infiltrated with continuous glass fibers, rolled, cooled, and shaped to prepare the FR/GF/PP prepregs. FR/GF/PP prepregs were cut according to the size of the mold; after cutting, the prepreg sheets were placed into the mold to be heated and melted (200 °C, 20 min), then quickly transferred into the molding machine at room temperature for rapid pressurization (25 °C, 7 MPa) and taken out after holding pressure for 10 min. Finally, they were cut into the required sizes for each test using waterjet cutting equipment. In order to reduce the effect of fiber orientation on performance, the FR/GF/PP laminates were unidirectional. The fiber content was 10% volume content.

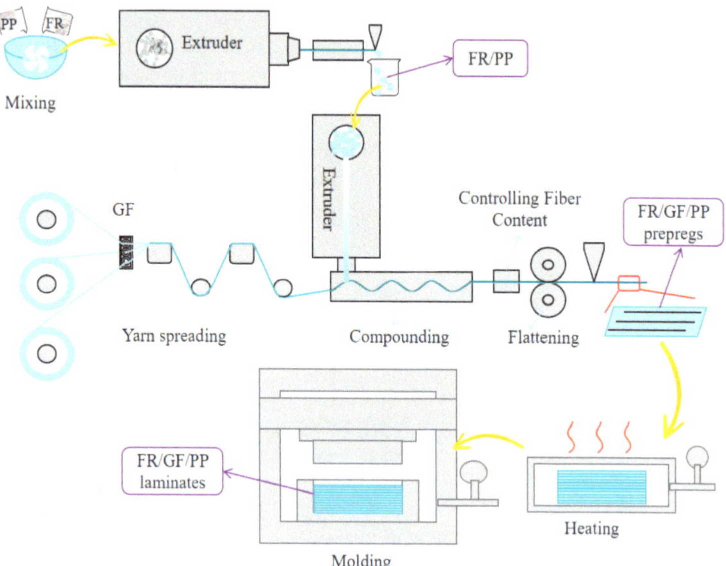

Figure 1. Schematic diagram of preparation process.

2.4. Characterization

2.4.1. Mechanical Properties

Mechanical property tests were performed on Changchun Kexin WDW-100 Universal Mechanical Testing Machine. According to the ASTM D638 standard [30], the tensile property was tested at the loading speed of 1 mm/min, with a specimen shape of Type I. According to the ASTM D7264 standard [31], a three-point bending performance test was carried out on each specimen at the loading speed of 2 mm/min, with a specimen span-to-thickness ratio of 32.

2.4.2. Dielectric Properties

The dielectric constant and dielectric loss of each specimen at X-band (8.2–12.4 GHz) electromagnetic waves were measured using the rectangular waveguide method on a Model AV3672C Vector Grid Analyzer (China Electronics Technology Group Corporation, Qingdao, China). The rectangular waveguide dielectric constant measurement method utilized the TE mode of electromagnetic waves transmitted from the waveguide to the material to be measured, and the amplitude and phase information of the reflected and transmitted signals were used to invert the dielectric constant of the material to be measured. The specimen size was 10.16 mm × 22.86 mm × 3.0 mm.

2.4.3. Limiting Oxygen Index (LOI)

The Limiting Oxygen Index of each specimen was measured using an Oxygen Indexer (Nanjing Jiangning Analytical Instrument Co., Ltd., Nanjing, China), according to the ASTM D2863 standard [32] test method. The LOI measured the minimum concentration of oxygen required for the flaming combustion of the material in a mixed oxygen-nitrogen gas stream. The LOI specimen size was 6.5 mm × 120 mm × 3.0 mm. The number of parallel specimens was about 15, and the experimental results were statistical results according to ASTM D2863.

2.4.4. Vertical Burning Test

Flame retardant rating was tested using a vertical combustion testing machine (YK-Y0142, YAOKE, Nanjing, China) according to the UL94 flame retardant rating (FRR) test method. The flame retardancy rating was used to evaluate the ability of a material to extinguish after ignition. The FRR specimen size was 13 mm × 120 mm × 3.0 mm. The number of parallel specimens was 10, and the experimental results were statistical results according to UL94.

2.4.5. Scanning Electron Microscope (SEM)

The microscopic morphology of each specimen was observed using a Focused Ion Beam-Scanning Electron Microscope (Helios G4 CX, ThermoFisher Scientific, Brno, Czech Republic). The observation position was the fracture part of the specimen after the bending test. The samples were sprayed gold and tested at an accelerating voltage of 15 kV. The magnification was 2000 times.

3. Results and Discussion

3.1. Mechanical Properties of GF/PP Modified with Different Flame Retardants

3.1.1. Tensile Properties of PP Modified with Different Flame Retardants

Among all the specimens, the polypropylene specimen without the flame-retardant addition is used as the blank control group. The tensile strength of the control group is 23.04 MPa and the tensile modulus is 1.60 GPa.

It is obvious from the test results in Figure 2 that the tensile strength and elongation at the break of FR/PP show a decreasing trend with the increase in flame-retardant content, while the tensile modulus increases gradually, whether it is DBDPE, MONO, MULTI or IFR flame retardant. There are differences in the effect of the type of flame retardant on the tensile properties of polypropylene.

When the amount of flame retardant added is lower (8%), the reduction in tensile strength of the four flame retardant-modified PP is not obvious. However, when the flame retardant additive amount is increased to 20.69%, the tensile strengths of MONO and MULTI-modified PP are 19.44 MPa and 19.69 MPa, which are significantly decreased compared to the control group without the flame retardant, which are decreased by 15.63% and 14.54%, respectively. The tensile strengths of DBDPE-modified PP are also decreased with the increase in the content of flame retardant; at flame-retardant additions of 8.00%, 14.81%, and 20.69%, the tensile strengths are 22.69 MPa, 21.48 MPa, and 20.60 MPa, respectively, which are decreased by 1.52%, 6.77%, and 10.59%, respectively, compared to the control specimens with no flame retardant added. IFR-modified PP shows stable tensile strength, and the flame-retardant content has no negative effect on the tensile strength of IFR-modified PP. Among the specimens with IFR as the added flame retardant, the tensile strengths are 22.56 MPa, 22.69 MPa, and 22.61 MPa at the flame-retardant additions of 8.00%, 14.81%, and 20.69%, respectively, and the tensile strengths of the IFR-modified PP show stable performance.

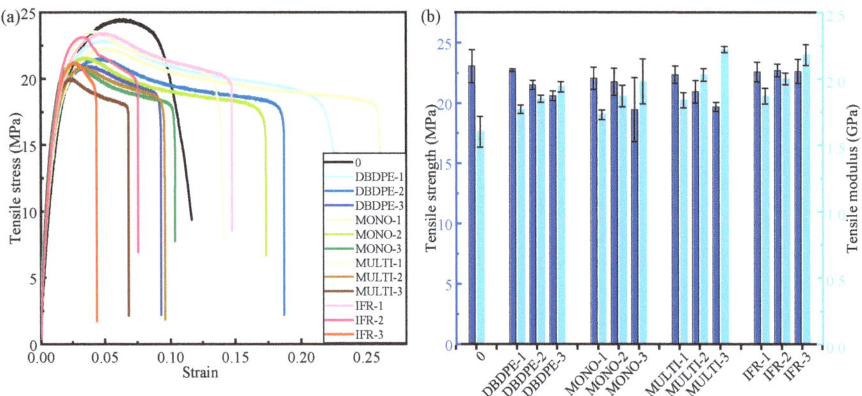

Figure 2. (**a**) Stress–strain curves, (**b**) tensile modulus and tensile strength of PP modified with different flame retardants.

The tensile modulus of the four flame retardant-modified PP increases with the increase in flame-retardant content. The tensile modulus of the polypropylene specimens with DBDPE, MONO, MULTI and IFR flame retardants reaches the maximum value of 1.94 GPa, 1.98 GPa, 2.22 GPa, and 2.18 GPa, respectively, when the flame-retardant content is 20.69%. Among them, the MULTI- and IFR-modified PP have the best effect in improving tensile modulus, which is higher than that of the control group by 38.75% and 36.25%, respectively.

The addition of flame retardants hinders the movement of the chain segments of PP, increases the stiffness, and increases the tensile modulus, but also introduces defects into the matrix, resulting in a decrease in tensile strength due to increased defects. Considering the results of combined tensile strength and tensile modulus, the tensile properties of IFR-modified PP have significant advantages. IFR has better compatibility with PP.

3.1.2. Bending Properties of GF/PP Composites Modified with Different Flame Retardants

The three-point bending properties of GF/PP composites modified with different flame retardants are presented in Figure 3. The bending strength of the blank control composite is 268.56 MPa and the bending modulus is 11.99 GPa.

The bending strength and bending modulus of each flame retardant-modified GF/PP show a decreasing trend with the increase in flame-retardant addition. Taking IFR/GF/PP as an example, when the flame retardant additive amount is 8.00%, the bending strength and bending modulus of IFR-modified PP are 269.4 MPa and 11.62 GPa, respectively, and the performance is almost unchanged compared with that of the blank control group, and even the average bending strength is slightly increased. When the flame retardant additive amount is increased to 20.69%, the bending strength and bending modulus decrease dramatically to 233.14 MPa and 9.54 GPa, respectively, which are 13.19% and 20.43% lower than the blank control group, respectively.

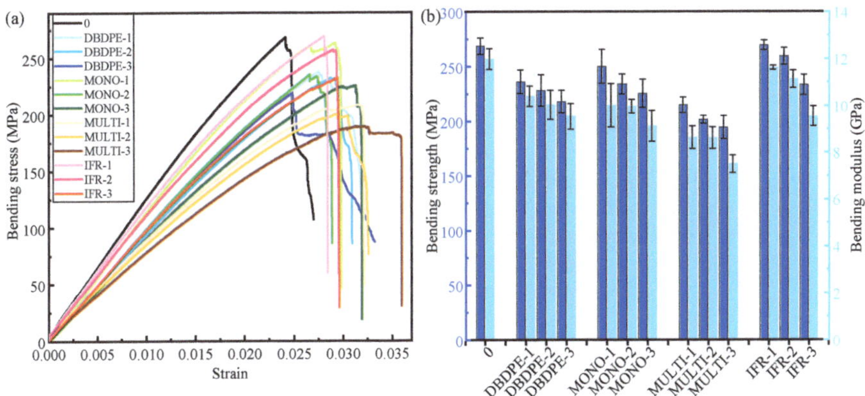

Figure 3. (**a**) Bending stress–strain curves, (**b**) bending modulus and bending strength of PP modified with different flame retardants.

Comparing the four flame retardants, the bending strength and bending modulus of the MULTI-modified GF/PP drop the most obviously, and the bending strength and bending modulus drop to 194.2 MPa and 7.5 GPa when the flame-retardant addition is increased to 20.69%, which is 27.69% and 37.45% lower than that of the blank control group, respectively. In Section 3.1.1, concerning the MONO and MULTI-modified PP, the tensile properties of the two are similar, but when they modify GF/PP there is a gap and the bending properties of the MULTI-modified GF/PP are significantly lower, which indicates that the MULTI-modified PP has poor composite properties with glass fibers and does not form an effective load transfer interface.

From Figure 4a–d, it can be seen that DBDPE-, MONO- and IFR-modified PP have a better compounding ability with fibers, and there is still a large amount of resin remaining on the surface of the fibers after they are detached, while MULTI-modified PP has very poor combining ability with fibers; the fiber surface is smooth and almost no MULTI-modified PP remains. From Figure 4d–f, it can be seen that with the increase in the flame-retardant content of the fiber, the surface residue of the resin decreases, and the IFR-modified PP and fiber composite ability decreases.

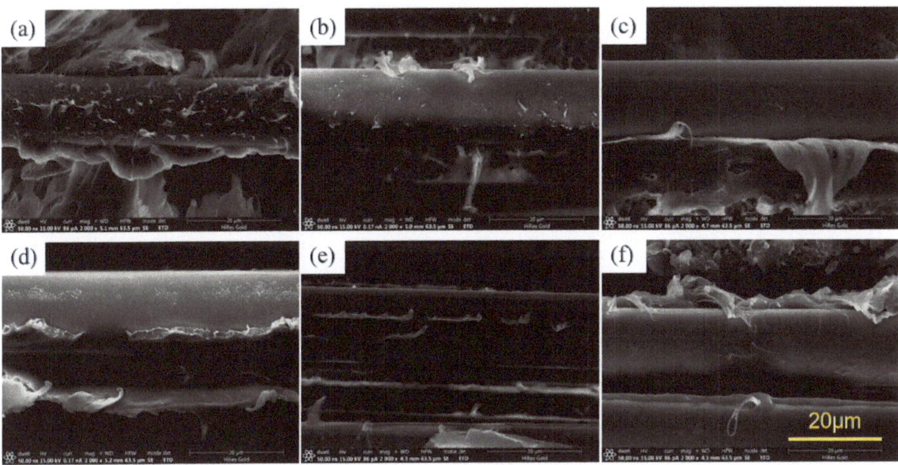

Figure 4. SEM images of GF/PP modified with different flame retardants, (**a**) DBDPE-1, (**b**) MONO-1, (**c**) MULTI-1, (**d**) IFR-1, (**e**) IFR-2, and (**f**) IFR-3.

In Section 3.1.1, the tensile modulus of FR/PP is enhanced with the increase in flame-retardant content, but the phenomenon where the bending modulus increases with the flame retardant does not occur in FR/GF/PP. This is due to the fact that the reinforcing phase in fiber-reinforced composites plays a major role in bearing [33], and the modulus of the matrix and the modulus of the reinforcing body are orders of magnitude different, so the increase in the modulus of the matrix does not play a large role in the overall bending modulus change in the fiber-reinforced composites. While the addition of flame retardant affects the interfacial properties between the matrix and the fiber, the interfacial phase has a particularly important role in the composite material, which is an extremely important microstructure of the composite material, and its structure and properties directly affect the performance of the composite material [34,35]. With the increase in flame-retardant content, the viscosity of matrix melt increases, resulting in a poorer effect of matrix infiltration of fibers, which leads to the interface not being effective in transferring loads and the mechanical properties being reduced. The information presented in Figure 3 shows that the effect of IFR flame retardants on the bending properties of FR/GF/PP is minimized at the same additive level.

3.2. Flame-Retardant Properties of GF/PP Composites Modified with Different Flame Retardants

3.2.1. LOI of GF/PP Composites Modified with Different Flame Retardants

The LOI data of different flame retardant-modified GF/PP composites are presented in Figure 5, from which it is relatively intuitive to observe that the LOI of flame retardant-modified GF/PP shows an increasing trend with the increase in the flame-retardant content, and the type of flame retardant has a great influence on the LOI of FR/GF/PP.

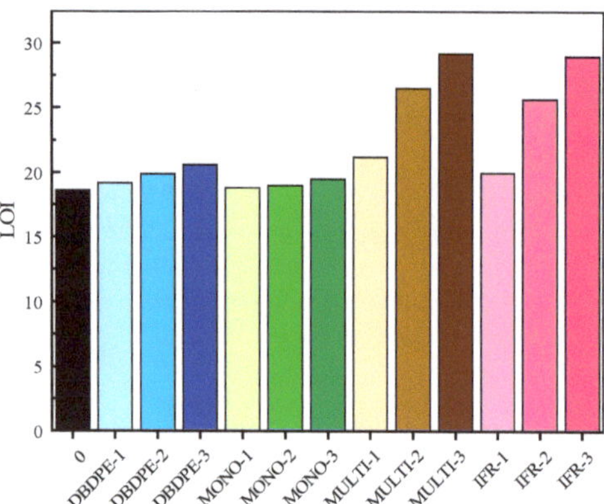

Figure 5. LOI of GF/PP composites modified with different flame retardants.

For the specimens with the addition of DBDPE and MONO flame retardants, the variation in the LOIs is not significant, the enhancement of flame retardant performance is limited, and the flame retardant effect is not ideal. The LOIs of the composite specimens with MULTI and IFR flame retardants increase significantly with the increase in flame-retardant addition. In the samples with MULTI flame retardant, the LOIs of the samples are 21.2, 26.5 and 29.2 at the content of 8.00%, 14.81% and 20.69% of flame retardant, respectively, which are 13.97%, 42.47% and 56.99% higher than that of the samples in the blank control group. In the samples with the IFR flame retardant, the LOIs of the samples are 20.0, 25.7 and 29.0 at the contents of 8.00%, 14.81% and 20.69% of flame retardant, respectively, which are 7.52%, 38.17% and 55.91% higher than those of the samples of the blank control group, respectively. MULTI and IFR are multi-component flame retardants and have higher LOI. Multi-component flame retardants can inhibit material combustion in more ways than a single component, which may result in a better LOI.

3.2.2. Flame Retardant Rating of GF/PP Composites Modified with Different Flame Retardants

The combustion process of the vertical combustion rating test of composite specimens with different flame retardants is presented in Figure 6.

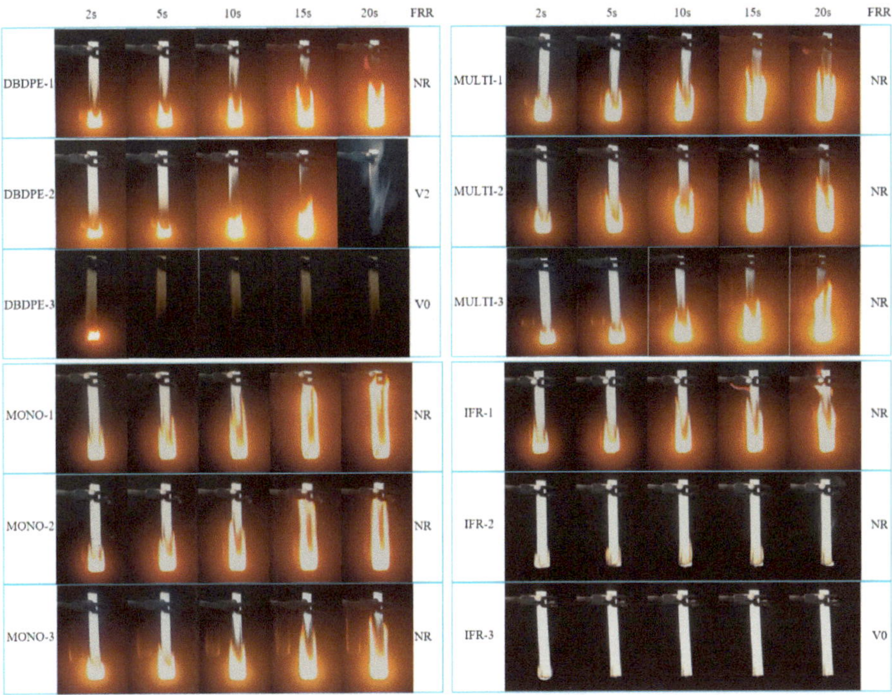

Figure 6. Test photos of vertical burning performance and FRR of GF/PP composites modified with different flame retardants.

At 8% flame-retardant content, all four types of flame retardant specimens appear to have flame extension and continuous combustion after 10 s of ignition time. The composite specimens with the MONO flame retardant have the fastest burning speed; the flame extended to the top of the specimen after 20 s of withdrawing from the ignition source, while the flame burned to the middle of the specimen for the other three types of specimens.

At 14.81% flame-retardant content, after 10 s of ignition time, the flame of the DBDPE-added specimen had a short extension, then the flame gradually weakened, and the combustion stopped at 20 s; it can be seen in the photo that a large amount of smoke is generated after the combustion has stopped. For the MULTI-added specimen, after the ignition time, the flame continued to expand, the combustion process continued, and after 20 s, the combustion flame was at the middle of the specimen. For the MONO-added specimen, after the fire source withdrew from the specimen, the flame expanded rapidly; during the observation time, the flame did not appear to weaken, and after 20 s the specimen combustion flame reached the specimen clamping position. For the IFR-added specimen, after 10 s of ignition time, the specimen ignited, but the flame was small and the flame expansion was very slow. After 20 s, although the flame was not completely extinguished, it seemed to be very weak.

At 14.81% flame-retardant content, specimens with DBDPE and IFR additions show a brief 1–3 s of burning after the ignition time, followed by immediate self-extinguishing.

The specimens with the MULTI and MONO additions continue to burn after the ignition time, with a significant expansion of the flame, and are still not extinguished after 20 s.

Among the four kinds of flame retardants, only the DBDPE-1 and DBDPE-2 specimens are accompanied with the phenomenon of resin molten droplets igniting the bottom skimming cotton during combustion. Finally, the FRRs are obtained for DBDPE-2-, DBDPE-3- and IFR-3-modified GF/PP specimens, and their FRRs are V2, V0 and V0, respectively.

3.3. Dielectric Properties of GF/PP Modified with Different Flame Retardants

The relation between the dielectric constant and the angle θ of the continuous fiber-reinforced composite [36] is shown in Equation (4). This rule does not change with the variation in flame-retardant content.

$$\varepsilon'_{11} = \varepsilon'_f v_f + \varepsilon'_m v_m \tag{1}$$

$$\varepsilon'_{22} = \frac{\varepsilon'^2_m + \varepsilon'_m \sqrt{v_f}\left(\bar{\varepsilon}'_{f2} - \varepsilon'_m\right)}{\varepsilon'_m\left(1 + v_f - \sqrt{v_f}\right) + \bar{\varepsilon}'_{f2}\left(\sqrt{v_f} - v_f\right)} \tag{2}$$

where

$$\bar{\varepsilon}'_{f2} = \varepsilon'_m + \frac{\pi}{4}\left(\varepsilon'_f - \varepsilon'_m\right) \tag{3}$$

$$\left[\varepsilon'_{ij}\right]^{\theta} = \varepsilon'_{11}cos^2\theta + \varepsilon'_{22}sin^2\theta \tag{4}$$

where ε'_{11} and ε'_{22} are the dielectric constants at angles of 0 and 90, respectively. ε'_f and ε'_m are the dielectric constants of the fiber and matrix, respectively. v_f and v_m are the volume contents of the fiber and matrix, respectively, and $\left[\varepsilon'_{ij}\right]^{\theta}$ is the dielectric constant when the angle is θ.

It can be concluded from Equation (4) that the dielectric properties of unidirectional continuous fiber composite panels decrease with the increase in the angle θ (Figure 7) between the fibers and the direction of the electric field, which is at its maximum at $\theta = 0°$ (parallel) and minimum at $\theta = 90°$ (vertical). So in this paper, the dielectric properties are tested for both $\theta = 0°$ and $\theta = 90°$ directions. The dielectric properties in each direction are shown in Figures 8 and 9.

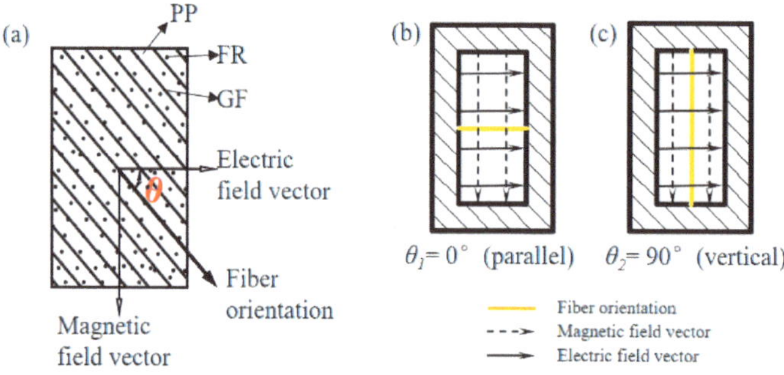

Figure 7. Schematic diagram of fiber direction and electric field direction, (a) θ position, (b) $\theta = 0°$, (c) $\theta = 90°$.

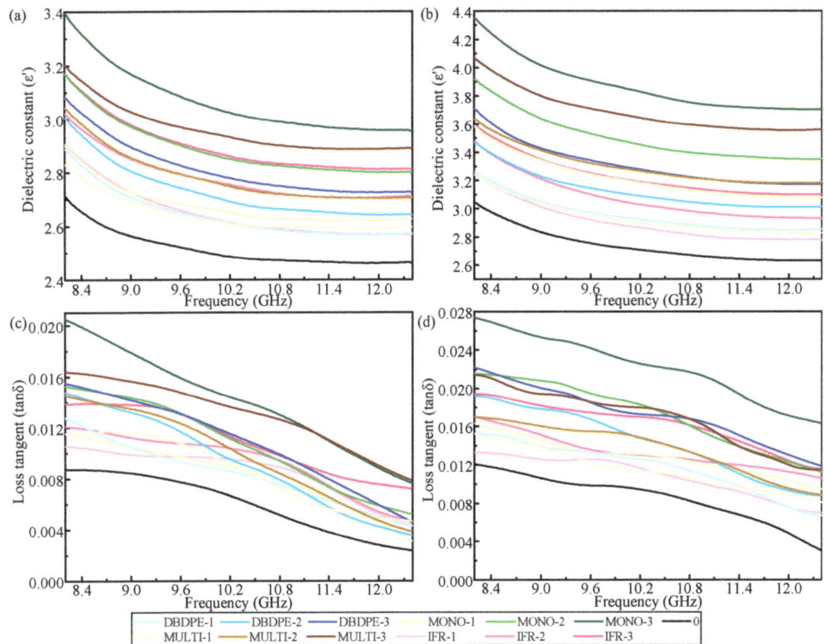

Figure 8. Dielectric properties versus frequency of FR/GF/PP samples, (**a**) dielectric constant at $\theta = 90°$, (**b**) dielectric constant at $\theta = 0°$, (**c**) dielectric loss tangent at $\theta = 90°$, and (**d**) dielectric loss tangent at $\theta = 0°$.

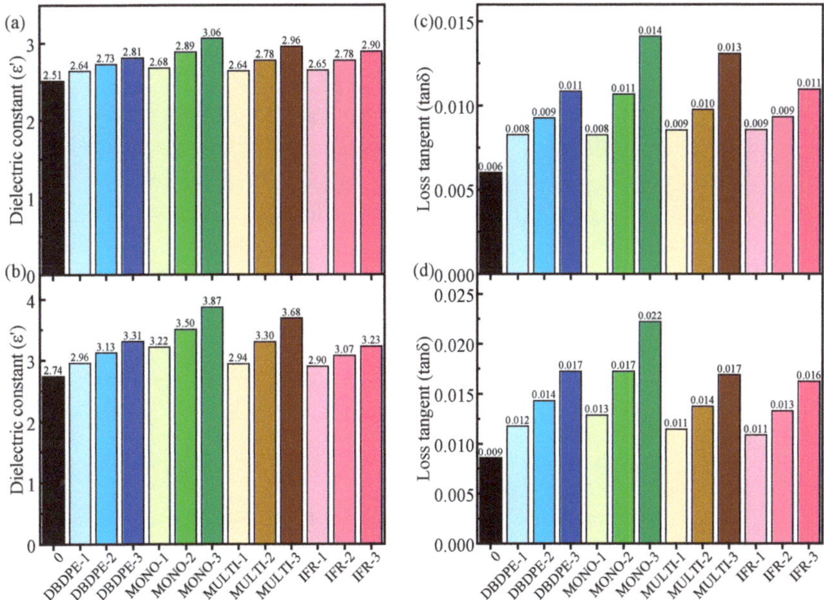

Figure 9. Average dielectric properties of FR/GF/PP samples in X-band, (**a**) dielectric constant at $\theta = 90°$, (**b**) dielectric constant at $\theta = 0°$, (**c**) dielectric loss tangent at $\theta = 90°$, and (**d**) dielectric loss tangent at $\theta = 0°$.

The dielectric constant and dielectric loss tangent of FR/GF/PP increase with the increase in flame-retardant content when the flame retardant type is the same. It can be hypothesized from this result that probably the dielectric properties of the flame retardant are higher than those of the PP matrix. As the flame-retardant content increases, the interfacial polarization between the resin and fibers is also enhanced, resulting in an increase in the total dielectric constant and dielectric loss tangent of the composite. The addition of large amounts of flame retardants has a very significant effect on the dielectric properties, especially the dielectric loss values, so it is necessary to reduce the flame-retardant content to maintain the low dielectric properties of polypropylene.

At the same flame-retardant content, the dielectric properties of different kinds of flame retardants have different extents of increase. When the flame-retardant content is increased to 20.69%, the parallel dielectric constants of GF/PP modified with four kinds of flame retardants (DBDPE, MONO, MULTI, IFR) increase to 3.31, 3.87, 3.68 and 3.23, respectively, which are 20.87%, 41.32%, 34.53% and 17.88% more than those of the blank group, and the vertical dielectric constants increase to 2.81, 3.06, 2.96 and 2.90, which are 11.88%, 21.92%, 17.70% and 15.23% more than those of the blank group, respectively.

When the flame-retardant content is increased to 20.69%, the parallel dielectric loss tangents of GF/PP modified with four kinds of flame retardants (DBDPE, MONO, MULTI, IFR) increase to 0.01723, 0.02219, 0.01686 and 0.01621, respectively, which are 101.29%, 159.23%, 96.96% and 89.37% more than those of the blank group, and the vertical dielectric loss tangents increase to 0.01083, 0.0141, 0.01307 and 0.01095, which are 79.60%, 133.83%, 116.75% and 81.59% more than those of the blank group, respectively.

Among the four different types of FR/GF/PP, DBDPE- and IFR-modified GF/PP have lower dielectric constants and dielectric loss tangents at the same flame-retardant content. This can be attributed to the fact that these two flame retardants may have lower polarizabilities themselves, and lower interfacial polarizations because of better interfaces (Figure 4).

The dielectric constant and dielectric loss tangent of each specimen show a decreasing trend with the increase in frequency, and this trend does not change with the addition of flame retardant type, which is due to the time needed for the polarization of the dielectric and because the polarization time of various polarization methods is different. With the frequency increase, the electric field change period becomes shorter, the internal polarization of the material gradually lags behind the electric field change, and part of the polarization will not work, so the dielectric properties are reduced.

3.4. Comprehensive Performance Evaluation

The above analysis can be used to briefly analyze the effects of flame retardant type and flame-retardant content on bending properties, flame-retardant properties and dielectric properties, respectively; for example, the increase in flame-retardant content is helpful for the improvement of flame retardant performance, but a large addition will reduce the mechanical properties and increase the dielectric properties. There is no one flame retardant that can obtain the optimal performance in all the properties; for example, IFR has the highest performance in mechanical properties, MULTI has a better performance in LOI, DBDPE and IFR have better performances in FRR, and DBDPE and IFR have lower and similar dielectric properties. Therefore, certain statistical rules are needed to select the most suitable flame retardant type and flame-retardant content value for the application in this paper from many combinations.

In order to obtain the optimal overall performance, gray relational analysis is applied to deal with the data of each specimen. Gray relational analysis can transform the multi-objective multi-response problem into a single gray relational comparison problem. The comprehensive performance of each specimen is compared by the final gray relational comparison.

The experiment data of each test are summarized in Table 2. In order to facilitate data processing, the flame retardant level V0 is assigned to 3, V1 is assigned to 2, V2 is assigned to 1, and no grade is assigned to 0. The dielectric constant and dielectric loss tangent are the average of the X-band test values in both directions.

Table 2. Experiment data of each sample.

Samples	Bending Strength (MPa)	Bending Modulus (GPa)	LOI (%)	FRR	Dielectric Constant	Dielectric Loss Tangent
0	268.56	11.99	18.6	0	2.6258	0.0073
DBDPE-1	236.00	10.39	19.2	0	2.7980	0.0100
DBDPE-2	228.09	10.03	19.9	1	2.9262	0.0118
DBDPE-3	217.84	9.53	20.6	3	3.0609	0.0140
MONO-1	249.98	10.00	18.8	0	2.9475	0.0106
MONO-2	233.88	9.96	19	0	3.1938	0.0139
MONO-3	225.56	9.12	19.5	0	3.4670	0.0181
MULTI-1	214.79	8.63	21.2	0	2.7896	0.0100
MULTI-2	201.35	8.61	26.5	0	3.0401	0.0117
MULTI-3	194.19	7.50	29.2	0	3.3210	0.0150
IFR-1	269.40	11.62	20	0	2.7740	0.0097
IFR-2	259.37	11.13	25.7	0	2.9272	0.0113
IFR-3	233.14	9.54	29	3	3.0620	0.0136

In order to eliminate the effect of magnitude, the data obtained from the experiments are homogenized. Among the objectives, bending strength, bending modulus, LOI and FRR are the larger-the-better objectives, which are homogenized using Equation (5); the dielectric constant and dielectric loss are the smaller-the-better objectives, which are processed using Equation (6). The values of each objective after homogenization are summarized in Table 3.

$$x_i(k) = \frac{y_i(k) - \min y_i(k)}{\max y_i(k) - \min y_i(k)} \tag{5}$$

$$x_i'(k) = \frac{\max y_i(k) - y_i(k)}{\max y_i(k) - \min y_i(k)} \tag{6}$$

Table 3. Homogenization date of each sample.

Samples	Bending Strength	Bending Modulus	LOI	FRR	Dielectric Constant	Dielectric Loss Tangent
0	0.99	1	0	0	1	1
DBDPE-1	0.56	0.64	0.06	0	0.80	0.75
DBDPE-2	0.45	0.56	0.12	0.33	0.64	0.59
DBDPE-3	0.31	0.45	0.19	1	0.48	0.38
MONO-1	0.74	0.56	0.02	0	0.62	0.70
MONO-2	0.53	0.55	0.04	0	0.32	0.39
MONO-3	0.42	0.36	0.08	0	0	0
MULTI-1	0.27	0.25	0.25	0	0.81	0.75
MULTI-2	0.10	0.25	0.75	0	0.51	0.59
MULTI-3	0	0	1	0	0.17	0.29
IFR-1	1	0.92	0.13	0	0.82	0.78
IFR-2	0.87	0.81	0.67	0	0.64	0.63
IFR-3	0.52	0.45	0.98	1	0.48	0.42

The homogenized data are processed using Equation (7) to obtain the gray correlation coefficient of each specimen in each performance. The gray correlation coefficient can reflect the degree of correlation between the actual value and the expected value, and the larger the gray correlation coefficient is, the closer it is to the optimization goal.

$$\xi_i(k) = \frac{\Delta\min + \zeta \times \Delta\max}{|x_0(k) - x_i(k)| + \zeta \times \Delta\max} \tag{7}$$

where

$$\Delta min = \min_i \min_k |x_0(k) - x_i(k)| \tag{8}$$

$$\Delta max = \max_i \max_k |x_0(k) - x_i(k)| \tag{9}$$

$x_0(k)$ is the reference data column and refers to the maximum value of each performance index after homogenization: here $x_0(k) = 1$. ζ is the resolution coefficient, $\zeta \in [0,1]$, and in this paper takes 0.5. $\xi_i(k)$ is called the correlation coefficient of x_i to x_0 with respect to the k indicator. The gray correlation coefficient of each performance is summarized in Table 4.

Table 4. Gray correlation coefficient of each sample.

Samples	Bending Strength	Bending Modulus	LOI	FRR	Dielectric Constant	Dielectric Loss Tangent
0	0.98	1.00	0.33	0.33	1.00	1.00
DBDPE-1	0.53	0.58	0.35	0.33	0.71	0.67
DBDPE-2	0.48	0.53	0.36	0.43	0.58	0.55
DBDPE-3	0.42	0.48	0.38	1.00	0.49	0.45
MONO-1	0.66	0.53	0.34	0.33	0.57	0.62
MONO-2	0.51	0.53	0.34	0.33	0.43	0.45
MONO-3	0.46	0.44	0.35	0.33	0.33	0.33
MULTI-1	0.41	0.40	0.40	0.33	0.72	0.67
MULTI-2	0.36	0.40	0.66	0.33	0.50	0.55
MULTI-3	0.33	0.33	1.00	0.33	0.38	0.41
IFR-1	1.00	0.86	0.37	0.33	0.74	0.69
IFR-2	0.79	0.72	0.60	0.33	0.58	0.58
IFR-3	0.51	0.48	0.96	1.00	0.49	0.46

The gray correlation for each specimen is the sum of the multiples of each performance weight and the corresponding gray correlation coefficient, which is calculated using Equation (10). R_i is the gray correlation of the ith specimen.

$$R_i = \sum_{k=1}^{n} \omega(k)\xi_i(k) \tag{10}$$

The evaluation of the weights for each objective is quite subjective, and different workers will have different weights assigned for different applications. The weights of each performance in this paper are obtained using the analytic hierarchy process (AHP) according to the set application. In the application scenario of this paper, the flame retardant performance is required to be the highest, which is required to reach V0 grade, followed by dielectric properties, which are required to be as low as possible. The bending properties (bending strength greater than 200 MPa, bending modulus greater than 9 GPa) can be achieved in most of the specimens; therefore, flame-retardant properties are considered the most important, followed by dielectric properties, and finally, mechanical properties.

AHP calculates weight vectors through building hierarchical models and constructing judgment matrices [37–39]. The hierarchical model designed is presented in Table 5.

Table 5. Hierarchical model.

A	B	C
	Flame-retardant properties (B1)	LOI (C1) FRR (C2)
Comprehensive performance (A1)	Dielectric properties (B2)	Dielectric constant (C3) Dielectric loss tangent (C4)
	Mechanical properties (B3)	Bending strength (C5) Bending modulus (C6)

The judgment matrix of the analytic hierarchy process is presented in Table 6.

Table 6. Judgment matrix.

A1	B1	B2	B3
B1	1	3	5
B2	1/3	1	3
B3	1/5	1/3	1

By using the judgment matrix, the weights of each performance are calculated according to Equation (11).

$$\omega_i = \frac{b_i}{\sum\limits_{k=1}^{n} b_k} \tag{11}$$

where

$$b_i = \left(\prod_{j=1}^{n} a_{ij} \right)^{\frac{1}{n}} \tag{12}$$

n is the number of indicator factors in each layer; a_{ij} is the relative importance of indicator i compared to factor j. The values of a_{ij} are within the judgment matrix (Table 6), with 1 indicating that factors i and j are equally important, 3 indicating that i is slightly more important than factor j, and 5 indicating that i is significantly more important than factor j.

Each performance of Layer C is considered equally important relative to Layer B. The final weights of each layer are presented in Table 7.

Table 7. Weights of each layer.

B	B~A	C	C~B	C~A
Flame-retardant properties (B1)	0.6370	LOI (C1)	0.5	0.3185
		FRR (C2)	0.5	0.3185
Dielectric properties (B2)	0.2583	Dielectric constant (C3)	0.5	0.1291
		Dielectric loss tangent (C4)	0.5	0.1291
Mechanical properties (B3)	0.1047	Bending strength (C5)	0.5	0.0524
		Bending modulus (C6)	0.5	0.0524

The weights from Table 8 and the gray correlation coefficients from Table 4 are brought into Equation (10) to calculate the gray correlation for each specimen.

Table 8. Gray correlation values for each specimen.

Samples	0	DBDPE-1	DBDPE-2	DBDPE-3	MONO-1	MONO-2	MONO-3
Gray correlation	0.57	0.45	0.45	0.61	0.43	0.38	0.35

Samples	MULTI-1	MULTI-2	MULTI-3	IFR-1	IFR-2	IFR-3
Gray correlation	0.45	0.49	0.56	0.5	0.53	0.8

As shown in Table 8, IFR-3 has the highest gray correlation value, and the excellent flame retardant performance of the IFR flame retardant has greatly improved its gray correlation value. IFR has a better compounding ability with GF/PP and has higher mechanical properties and lower dielectric properties. The second is DBDPE-3, which is better in flame retardancy; the GF/PP with DBDPE flame retardant has the lowest dielectric properties, but its mechanical properties are poor, resulting in the overall evaluation of the second to IFR. Due to the addition of flame retardants, in the other combinations the mechanical properties deteriorate, the dielectric properties increase and the flame retardant performance is mediocre, resulting in gray correlation values even lower than the blank group of specimens. Therefore, in the application scenario set in this paper, the IFR flame retardant with 20.69% flame-retardant content is selected. Gray relational analysis is generally used for process parameter optimization, but in this paper it is applied for formulation selection. It provides ideas for multi-objective multi-response material formulation selection.

4. Conclusions

When the flame retardant types are the same, as the most important role of adding flame retardants, the flame-retardant properties of FR/GF/PP increase with the increase in flame-retardant content. However, with the increase in flame-retardant content, the mechanical properties of FR/GF/PP will be reduced and the dielectric properties will be elevated.

In addition to the flame-retardant content, the flame retardant type also has a significant effect on the properties of flame retardant-modified GF/PP. The effect of the flame retardant type on various properties is inconsistent. This is determined by their respective elemental composition and structure. Among the four flame retardant-modified GF/PP, IFR has the highest performance in mechanical properties, MULTI has a better performance in LOI, DBDPE and IFR have better performances in FRR, and DBDPE and IFR have lower and similar dielectric properties. Thus, it is necessary to find the most suitable flame retardant type and flame-retardant content value for the application scenario through certain statistical rules.

In this paper, gray relational analysis is used for formulation selection. The analytic hierarchy process is used to assign the weights of each performance target in combination with the set application occasions, and finally, the flame retardant type and flame-retardant content that optimize the comprehensive performance of FR/GF/PP, namely IFR-3, are obtained.

Author Contributions: Methodology, B.Z.; Resources, B.Z. and G.Q.; Data curation, J.L.; Writing—original draft, J.L.; Writing—review & editing, J.L.; Supervision, Y.S. and B.Z.; Project administration, Y.S. and G.Q. All authors have read and agreed to the published version of the manuscript.

Funding: This research was funded by the National Natural Science Foundation of China (Grant No. 12302162).

Institutional Review Board Statement: Not applicable.

Data Availability Statement: Data are contained within the article.

References

1. Fan, X.; He, J.; Mu, J.; Qian, J.; Zhang, N.; Yang, C.; Hou, X.; Geng, W.; Wang, X.; Chou, X. Triboelectric-electromagnetic hybrid nanogenerator driven by wind for self-powered wireless transmission in Internet of Things and self-powered wind speed sensor. *Nano Energy* **2020**, *68*, 104319. [CrossRef]
2. Khan, S.; Ali, H.; Khalily, M.; Shah, S.U.; Kazim, J.U.; Ali, H.; Tanougast, C. Miniaturization of Dielectric Resonator Antenna by using Artificial Magnetic Conductor surface. *IEEE Access* **2020**, *8*, 68548–68558. [CrossRef]
3. Di Paola, C.; Zhao, K.; Zhang, S.; Pedersen, G.F. SIW Multibeam Antenna Array at 30 GHz for 5G Mobile Devices. *IEEE Access* **2019**, *7*, 73157–73164. [CrossRef]
4. Qiao, Z.; Wang, Z.; Loh, T.H.; Gao, S.; Miao, J. A Compact Minimally Invasive Antenna for OTA Testing. *IEEE Antennas Wirel. Propag. Lett.* **2019**, *18*, 1381–1385. [CrossRef]
5. Watanabe, A.O.; Tehrani, B.K.; Ogawa, T.; Raj, P.M.; Tentzeris, M.M.; Tummala, R.R. Ultralow-Loss Substrate-Integrated Waveguides in Glass-Based Substrates for Millimeter-Wave Applications. *IEEE Trans. Compon. Packag. Manuf. Technol.* **2020**, *10*, 531–533. [CrossRef]
6. Wang, C.H.; Rose, L.R.F. Wave reflection and transmission in beams containing delamination and inhomogeneity. *J. Sound Vib.* **2003**, *264*, 851–872. [CrossRef]

7. Du, W.; Zhou, Y.; Yao, Z.; Huang, Y.; He, C.; Zhang, L.; He, Y.; Zhu, L.; Xu, X. Active broadband terahertz wave impedance matching based on optically doped graphene-silicon heterojunction. *Nanotechnology* **2019**, *30*, 195705. [CrossRef] [PubMed]

8. Man, Z.; Li, P.; Zhou, D.; Wang, Y.; Liang, X.; Zang, R.; Li, P.; Zuo, Y.; Lam, Y.M.; Wang, G. Two Birds with One Stone: FeS2@C Yolk–Shell Composite for High-Performance Sodium-Ion Energy Storage and Electromagnetic Wave Absorption. *Nano Lett.* **2020**, *20*, 3769–3777. [CrossRef] [PubMed]

9. Guo, W.; Mias, C.; Farsad, N.; Wu, J.L. Molecular Versus Electromagnetic Wave Propagation Loss in Macro-Scale Environments. *IEEE Trans. Mol. Biol. Multi-Scale Commun.* **2015**, *1*, 18–25. [CrossRef]

10. Sun, B.H.; Wang, X.H.; Gao, Y. Study on the factors affecting the transmitting property of magnetic wave through glass/unsaturated polyester resin composites. *Fiber Compos.* **2002**, *2*, 13–16.

11. Huang, C.; Li, J.; Xie, G.; Han, F.; Huang, D.; Zhang, F.; Zhang, B.; Zhang, G.; Sun, R.; Wong, C.P. Low-Dielectric Constant and Low-Temperature Curable Polyimide/POSS Nanocomposites. *Macromol. Mater. Eng.* **2019**, *304*, 1900505. [CrossRef]

12. Vaidya, U.K.; Chawla, K.K. Processing of fibre reinforced thermoplastic composites. *Int. Mater. Rev.* **2008**, *53*, 185–218. [CrossRef]

13. Minchenkov, K.; Vedernikov, A.; Safonov, A.; Akhatov, I. Thermoplastic Pultrusion: A Review. *Polymers* **2021**, *13*, 180. [CrossRef]

14. Zhuo, P.; Li, S.; Ashcroft, I.A.; Jones, A.I. Material extrusion additive manufacturing of continuous fibre reinforced polymer matrix composites: A review and outlook. *Compos. Part B Eng.* **2021**, *224*, 109143. [CrossRef]

15. Khouaja, A.; Koubaa, A.; Ben Daly, H. Dielectric properties and thermal stability of cellulose high-density polyethylene bio-based composites. *Ind. Crops Prod.* **2021**, *171*, 113928. [CrossRef]

16. Sridhar, M.; Vasudeva Setty, R.N.; Johns, J. Electrical Properties of Bamboo Fiber Reinforced Polypropylene Composite: Effect of Coupling Agent. *J. Nat. Fibers* **2022**, *19*, 5076–5087. [CrossRef]

17. Xiong, J.; Wang, X.; Zhang, X.; Xie, Y.; Lu, J.; Zhang, Z. How the biaxially stretching mode influence dielectric and energy storage properties of polypropylene films. *J. Appl. Polym. Sci.* **2021**, *138*, 50029. [CrossRef]

18. Du, B.X.; Xu, H.; Li, J.; Li, Z. Space charge behaviors of PP/POE/ZnO nanocomposites for HVDC cables. *IEEE Trans. Dielectr. Electr. Insul.* **2016**, *23*, 3165–3174. [CrossRef]

19. Gibson, A.G.; Torres, M.O.; Browne, T.N.; Feih, S.; Mouritz, A.P. High temperature and fire behaviour of continuous glass fibre/polypropylene laminates. *Compos. Part A Appl. Sci. Manuf.* **2010**, *41*, 1219–1231. [CrossRef]

20. Zhao, W.; Cheng, Y.; Li, Z.; Li, X.; Zhang, Z. Improvement in fire-retardant properties of polypropylene filled with intumescent flame retardants, using flower-like nickel cobaltate as synergist. *J. Mater. Sci.* **2021**, *56*, 2702–2716. [CrossRef]

21. Chen, B.; Gao, W.; Shen, J.; Guo, S. The multilayered distribution of intumescent flame retardants and its influence on the fire and mechanical properties of polypropylene. *Compos. Sci. Technol.* **2014**, *93*, 54–60. [CrossRef]

22. Chen, H.; Wang, J.; Ni, A.; Ding, A.; Sun, Z.; Han, X. Effect of novel intumescent flame retardant on mechanical and flame retardant properties of continuous glass fibre reinforced polypropylene composites. *Compos. Struct.* **2018**, *203*, 894–902. [CrossRef]

23. Zhou, Y.; He, W.; Wu, Y.; Xu, D.; Chen, X.; He, M.; Guo, J. Influence of thermo-oxidative aging on flame retardancy, thermal stability, and mechanical properties of long glass fiber–reinforced polypropylene composites filled with organic montmorillonite and intumescent flame retardant. *J. Fire Sci.* **2019**, *37*, 176–189. [CrossRef]

24. Xu, J.; Li, K.; Deng, H.; Lv, S.; Fang, P.; Liu, H.; Shao, Q.; Guo, Z. Preparation of MCA-SiO$_2$ and Its Flame Retardant Effects on Glass Fiber Reinforced Polypropylene. *Fibers Polym.* **2019**, *20*, 120–128. [CrossRef]

25. Zhou, D.; He, W.; Wang, N.; Chen, X.; Guo, J.; Ci, S. Effect of thermo-oxidative aging on the mechanical and flame retardant properties of long glass fiber-reinforced polypropylene composites filled with red phosphorus. *Polym. Compos.* **2018**, *39*, 2634–2642. [CrossRef]

26. Li, J.; Sun, Y.; Zhang, B.; Qi, G. Mechanical, Flame-Retardant and Dielectric Properties of Intumescent Flame Retardant/Glass Fiber-Reinforced Polypropylene through a Novel Dispersed Distribution Mode. *Polymers* **2024**, *16*, 1341. [CrossRef]

27. Lv, H.; Chen, X.; Wang, X.; Zeng, X.; Ma, Y. A novel study on a micromixer with Cantor fractal obstacle through grey relational analysis. *Int. J. Heat Mass Transf.* **2022**, *183*, 122159. [CrossRef]

28. Hashemi, S.H.; Karimi, A.; Tavana, M. An integrated green supplier selection approach with analytic network process and improved Grey relational analysis. *Int. J. Prod. Econ.* **2015**, *159*, 178–191. [CrossRef]

29. Yu, X.; Shi, G.; Yang, X.; Gao, W. Research on tribological performance of textured thrust bearing using gray relational degree and improved multi-objective water circulation algorithm. *Surf. Topogr. Metrol. Prop.* **2024**, *12*, 25003. [CrossRef]

30. *ASTM D638*; Standard Test Method for Tensile Properties of Plastics. American Society for Testing and Materials: West Conshohocken, PA, USA, 2022.

31. *ASTM D7264*; Standard Test Method for Flexural Properties of Polymer Matrix Composite Materials. American Society for Testing and Materials: West Conshohocken, PA, USA, 2021.

32. *ASTM D2863*; Standard Test Method for Measuring the Minimum Oxygen Concentration to Support Candle-like Combustion of Plastics (Oxygen Index). American Society for Testing and Materials: West Conshohocken, PA, USA, 2019.

33. Meireman, T.; Daelemans, L.; Rijckaert, S.; Rahier, H.; Van Paepegem, W.; De Clerck, K. Delamination resistant composites by interleaving bio-based long-chain polyamide nanofibers through optimal control of fiber diameter and fiber morphology. *Compos. Sci. Technol.* **2020**, *193*, 108126. [CrossRef]

34. Huan, X.; Shi, K.; Yan, J.; Lin, S.; Li, Y.; Jia, X.; Yang, X. High performance epoxy composites prepared using recycled short carbon fiber with enhanced dispersibility and interfacial bonding through polydopamine surface-modification. *Compos. Part B Eng.* **2020**, *193*, 107987. [CrossRef]

35. Samyn, P. Engineering the Cellulose Fiber Interface in a Polymer Composite by Mussel-Inspired Adhesive Nanoparticles with Intrinsic Stress-Sensitive Responsivity. *ACS Appl. Mater. Interfaces* **2020**, *12*, 28819–28830. [CrossRef]
36. Chin, W.S.; Lee, D.G. Binary mixture rule for predicting the dielectric properties of unidirectional E-glass/epoxy composite. *Compos. Struct.* **2006**, *74*, 153–162. [CrossRef]
37. Mansor, M.R.; Sapuan, S.M.; Zainudin, E.S.; Nuraini, A.A.; Hambali, A. Hybrid natural and glass fibers reinforced polymer composites material selection using Analytical Hierarchy Process for automotive brake lever design. *Mater. Des.* **2013**, *51*, 484–492. [CrossRef]
38. Venkata Rao, R. Evaluation of metal stamping layouts using an analytic hierarchy process method. *J. Mater. Process. Technol.* **2004**, *152*, 71–76. [CrossRef]
39. Lee, D.; Lee, D.; Lee, M.; Kim, M.; Kim, T. Analytic Hierarchy Process-Based Construction Material Selection for Performance Improvement of Building Construction: The Case of a Concrete System Form. *Materials* **2020**, *13*, 1738. [CrossRef]

Article

A Novel Method of Improving the Mechanical Properties of Propellant Using Energetic Thermoplastic Elastomers with Bonding Groups

Shixiong Sun [1,2], Haoyu Liu [1], Yang Wang [1], Wenhao Du [1,2], Benbo Zhao [1,2,*] and Yunjun Luo [3,*]

[1] School of Chemistry and Chemical Engineering, North University of China, Taiyuan 030051, China; sunshixiong1989@126.com (S.S.); liuhaoyu8091@163.com (H.L.); 15227110251@163.com (Y.W.); wenhao_du@outlook.com (W.D.)

[2] Dezhou Industrial Technology Research Institute of North University of China, Dezhou 253034, China

[3] School of Materials Science and Technology, Beijing Institute of Technology, Beijing 100081, China

* Correspondence: zhaobenbo@163.com (B.Z.); yjluo@bit.edu.cn (Y.L.)

Abstract: The relatively poor mechanical properties of extruded modified double base (EMDB) propellants limit their range of applications. To overcome these drawbacks, a novel method was proposed to introduce glycidyl azide polymer-based energetic thermoplastic elastomers (GAP-ETPE) with bonding groups into the propellant adhesive. The influence of the molecular structure of three kinds of elastomers on the mechanical properties of the resultant propellant was analyzed. It was found that the mechanical properties of the propellant with 3% CBA-ETPE (a type of GAP-ETPE that features chain extensions using N-(2-Cyanoethyl) diethanolamine and 1,4-butanediol) were improved at both 50 °C and −40 °C compared to a control propellant without GAP-ETPE. The elongation and impact strength of the propellant at −40 °C were 7.49% and 6.58 MPa, respectively, while the impact strength and maximum tensile strength of the propellant at 50 °C reached 21.1 MPa and 1.19 MPa, respectively. In addition, all three types of GAP-ETPE improved the safety of EMDB propellants. The friction sensitivity of the propellant with 3% CBA-ETPE was found to be 0%, and its characteristic drop height H_{50} was found to be 39.0 cm; 126% higher than the traditional EMDB propellant. These results provide guidance for studies aiming to optimize the performance of EMDB propellants.

Keywords: EMDB propellant; GAP-ETPE; interface properties; mechanical performance

Citation: Sun, S.; Liu, H.; Wang, Y.; Du, W.; Zhao, B.; Luo, Y. A Novel Method of Improving the Mechanical Properties of Propellant Using Energetic Thermoplastic Elastomers with Bonding Groups. *Polymers* **2024**, *16*, 792. https://doi.org/10.3390/polym16060792

Academic Editor: Andrea Sorrentino

Received: 22 February 2024
Revised: 6 March 2024
Accepted: 11 March 2024
Published: 13 March 2024

1. Introduction

Extruded modified double-base (EMDB) propellants with aluminum (Al) and hexogen (RDX) particles are one of the commonly used solid propellants for tactical rockets due their numerous advantages, including low cost, high production efficiency, mature process, and consistency between batches [1]. However, these propellants usually have poor mechanical properties, especially with regard to their toughness at low temperatures. In addition, the elongation of the propellants is only about 3% and their impact strength is only about 3 MPa at −40 °C when the solid particle content of the propellant is as high as 55% [2,3]. These are primarily due to the structure of the adhesive of EMDB propellants. Nitrocellulose (NC) is a rigid polymer used as the adhesive skeleton of these propellants; this gives the material its low toughness due to the nitrocellulose/nitroglycerin (NC/NG) adhesive system [4,5]. However, the interface character between the double base adhesive and RDX is poor, resulting in low adhesion. This weak interface is another weakness in the EMDB propellant that further reduces its mechanical properties [1]. Furthermore, NG is very sensitive, making the preparation and use of the EMDB propellant more dangerous. These drawbacks limit the use of EMDB propellants in a variety of applications.

Two methods can be used to improve the mechanical properties of EMDB propellants. First, the use of NG can be substituted with other nitrate esters such as diethyleneglycol dinitrate (DEGDN), triethylenglycol dinitrate (TEGDN), trimethylolethane trinitrate

(TMETN), Butyl Nitroxyethyl Nitramine (Bu-NENA) [4,6,7]. Indeed, some nitrate esters with lower sensitivity can plasticize NC better than NG. This would not only improve the mechanical properties of the propellant but also reduce its sensitivity. This is a common strategy in the regulation of the mechanical properties of propellants. However, those nitrate esters tend to reduce the high-temperature strength of EMDB propellants. A second method of improving the interface performance between NC/NG adhesive and RDX is the surface modification [8–10] of RDX with techniques such as spheroidization and surface coating. Although this process generally improves the mechanical properties of the target propellants, it is relatively complex. We previously introduced Bu-NENA, an insensitive nitrate ester with improved flexibility, into an EMDB propellant named R3 [11]. The elongation of the R3 propellant at $-40\,^{\circ}$C increased by a factor of 2 from 3.54% to 7.09%. In addition, the friction sensitivity of R3 dropped from 46% to 0%, while its H_{50} increased by 87.2% from 17.2 cm to 32.2 cm. In summary, the low-temperature mechanical properties and sensitivity of the EMDB propellant were improved significantly. Further improvements in mechanical properties could be achieved if the interface performance between the adhesive and RDX could be optimized.

Glycidyl azide polymer-based energetic thermoplastic elastomer (GAP-ETPE) is a linear polymer; propellants based on this polymer are known as green solid propellants due to the so-called 3R characteristics (recycle, recover, reuse) [12,13]. In our previous study, a series of GAP-ETPE with binding groups were synthesized [14]. Some of these materials were used as propellant adhesives when exploring the use of green or high-strength propellants [15]. An analysis of the phase structure revealed that NC molecules could diffuse into GAP-ETPE molecules due to the plasticization of Bu-NENA. Consequently, this suggests that GAP-ETPE with bonding groups could be dispersed into NC/Bu-NENA adhesives at the molecular level, improving its interface compatibility; this would result in improved mechanical properties in EMDB propellants with high-solid contents.

This study proposes a novel method of toughening the EMDB propellant based on the advantages of GAP-ETPE with bonding groups. The influence of GAP-ETPE on the tensile and impact properties of the EMDB propellant was studied in the context of its molecular structure. It was found that the mechanical properties of the E3C propellant (with 3% CBA-ETPE) could be improved at $50\,^{\circ}$C and $-40\,^{\circ}$C simultaneously. These results may provide insights into the improved performance of the EMDB propellant and the structural design of adhesives for a high-performance solid propellant.

Table 1 provides a list of abbreviations used in this manuscript.

Table 1. Abbreviations used in this work.

Abbreviation	Meaning
EMDB	Extruded modified double base
GAP	Glycidyl azide polymer
ETPE	Energetic thermoplastic elastomers
GAP-ETPE	Energetic thermoplastic elastomers based on glycidyl azide polymer
BDO	1,4-butanediol
DBM	Diethyl Bis(hydroxymethyl)malonate
CBA	N-(2-Cyanoethyl) diethanolamine
BDO-ETPE	GAP-ETPE with BDO as chain extender
DBM-ETPE	GAP-ETPE with chain extended by BDO and DBM (mass ratio 3:1)
CBA-ETPE	GAP-ETPE with chain extended by BDO and CBA (mass ratio 1:1)
R3	EMDB propellant without GAP-ETPE as a control
ExB	Propellant based on BDO-ETPE, where x refers to BDO-ETPE content
ExD	Propellant based on DBM-ETPE, where x refers to BDO-ETPE content
ExC	Propellant based on CBA-ETPE, where x refers to BDO-ETPE content
Ad-R3	The adhesive of R3 propellant
Ad-E3B	The adhesive of E3B propellant
Ad-E3D	The adhesive of E3D propellant
Ad-E3C	The adhesive of E3C propellant

2. Materials and Methods

2.1. Experimental Materials

Three kinds of GAP-ETPE with 30% hard-segment content were synthesized in the laboratory. Here, BDO-ETPE refers to an elastomer with BDO as a chain extender, while CBA-ETPE refers to an elastomer with its chain extended by BDO and CBA in a mass ratio of 1:1. DBM-ETPE refers to an elastomer with its chain extended by BDO and DBM in a mass ratio of 3:1. Molecular structure diagrams of these three types of elastomers are presented in Figure 1. The synthesis routes of these three types of GAP-ETPE are presented in Figure S1. The basic physical properties of these elastomers are shown in Table 2.

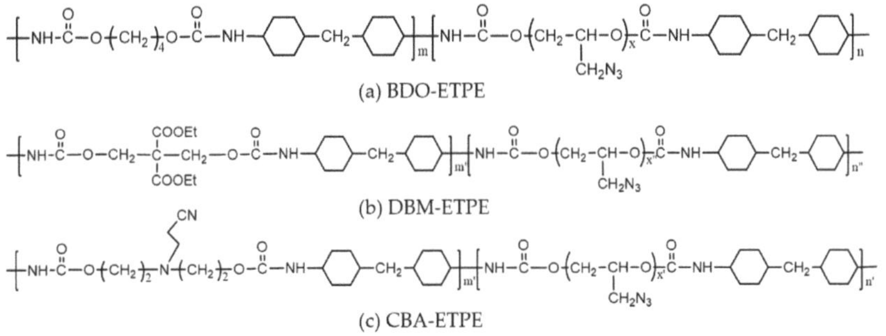

(a) BDO-ETPE

(b) DBM-ETPE

(c) CBA-ETPE

Figure 1. The molecular structure diagram of three kinds of GAP-ETPE: (**a**) BDO-ETPE; (**b**) DBM-ETPE; (**c**) CBA-ETPE.

Table 2. The physical properties of the three kinds of GAP-ETPE with 30% hard-segment content used in this study.

Property	BDO-ETPE	CBA-ETPE	DBM-ETPE
Average molecular weight: $\overline{M}_n$/g mol^{-1}	30,300	29,000	30,100
Density: ρ/g cm^{-3}	1.22	1.22	1.22
Glass transition temperature: T_g/°C	−39.4	−38.2	−39.0

The NC used in this study had an N content of 12.0% and an average molecular weight of about 80,000; it was obtained from the Shanxi Xingan Chemical Plant, Taiyuan City, China. Bu-NENA is a light-yellow oily liquid with a melting point of −28.0 °C, a density of 1.21 g/cm^3, and a heat of formation of 249 kJ/mol; it was obtained from the Liming Chemical Institute, Luoyang City, China. A 72-μm hexogen (RDX) was obtained from Gansu Silver Light Chemical Industry Group Co., Ltd., Baiyin City, China. A 3-μm spherical Al was obtained from Changyuan Mingyu Aluminium Industry Co., Ltd., Xinxiang City, China.

2.2. Sample Preparation

EMDB propellants with GAP-ETPE incorporated into their adhesive were prepared using a traditional solvent-free method. An EMDB propellant without GAP-ETPE (R3) was also prepared as a control. The detailed chemical composition of the prepared propellants is shown in Table 3. It should be noted that the R3 propellant was the previously optimized formula on which the series of three EMDB propellants with GAP-ETPE (ExB, ExD, and ExC) were prepared. The NC/Bu-NENA adhesive decreased with increasing GAP-ETPE content. The mass ratio of NC/Bu-NENA was held constant at 43/57, based on the results of previous experiments [11].

Table 3. Composition of EMDB propellants with different GAP-ETPE content.

Sample	NC/%	Bu-NENA/%	GAP-ETPE/%	Al/%	RDX/%	Others/%
R3	23.6	17.8	--	6	49	3.6
E1B	23.0	17.4	1	6	49	3.6
E2B	22.4	17.0	2	6	49	3.6
E3B	21.9	16.5	3	6	49	3.6
E4B	21.3	16.1	4	6	49	3.6
E5B	20.8	15.6	5	6	49	3.6
E1C	23.0	17.4	1	6	49	3.6
E2C	22.4	17.0	2	6	49	3.6
E3C	21.9	16.5	3	6	49	3.6
E4C	21.3	16.1	4	6	49	3.6
E5C	20.8	15.6	5	6	49	3.6
E1D	23.0	17.4	1	6	49	3.6
E2D	22.4	17.0	2	6	49	3.6
E3D	21.9	16.5	3	6	49	3.6
E4D	21.3	16.1	4	6	49	3.6
E5D	20.8	15.6	5	6	49	3.6

2.3. Experimental Instruments and Test Conditions

2.3.1. Interface Property Test

The interface characteristics of the adhesives were measured with an OCA contact angle analyzer (Dataphysics Co., Stuttgart, Germany). The testing solvents included *N,N*-dimethylformamide (DMF), ethylene glycol, and diiodomethane.

2.3.2. Tensile Property Test

The EMDB propellants were cut into dumbbell-shaped test specimens. The tensile properties of the propellants were measured using the AGS-J Electronic Universal Testing Machine (Shimadzu Corporation, Kyoto, Japan) according to the China Military Standard GJB 770B-2005 413.1 [16]. The test conditions were as follows: temperature $-40\ °C$, $20\ °C$, and $50\ °C$ at a tensile rate of 10 mm/min. The testing machine was equipped with a high–low-temperature test box through which the ambient temperature could be regulated.

2.3.3. Impact Property Test

The impact resistance of the propellant samples was measured using a TCJ liquid crystal pendulum impact-testing machine from Jilin Taihe Testing Machine Co., Ltd. (Changchun, China). The test temperatures were $-40\ °C$, $20\ °C$, and $50\ °C$; the pendulum energy was 4 J; the size of the sample was 4 mm × 6 mm × 60 mm, and the support span was 40 mm.

2.3.4. Sensitivity Measurements

The friction sensitivity of the propellants was measured according to China Military Standard GJB 770B-2005 601.2 [16] using a pendulum friction apparatus (Beijing nachen Technology Co., Ltd., Beijing, China). The conditions were as follows: a pendulum weight of 1.5 kg; a swaying angle of 66°; a pressure of 2.45 MPa; and a sample mass of 20 ± 1 mg. The impact sensitivity was measured according to China Military Standard GJB 770B-2005 602.1 [16] using a drop-hammer apparatus (Beijing nachen Technology Co., Ltd., Beijing, China) using an up-and-down method. The conditions were as follows: a sample mass of 30 mg and a hammer weight of 2 kg.

3. Results

3.1. Interface Characteristics of the NC/NENA/ETPE Adhesive

Solid propellant is a composite material comprising adhesive as the continuous phase and solid filler as the dispersed phase. Its mechanical properties are related not only to the mechanical properties of the adhesive itself but also to the interaction between the adhesive and solid filler [17,18]. The characteristics of the interface between the adhesive and solid filler are an important factor that affects the mechanical properties of the propellant. The surface energy of the adhesive is most commonly obtained using the contact angle method based on Young's equation [19–21]. In this study, the adhesion work between the adhesive and RDX was calculated using this method.

The adhesive systems were prepared as thin films. Diiodomethane, N,N-dimethylformamide, and ethylene glycol were used as reference liquids to test the contact angles of the four adhesive films at room temperature. The results of these tests are presented in Table 4. The interfacial tension and adhesion work between the Bu-NENA/NC/GAP-ETPE adhesive system and RDX were calculated [19–21] as shown in Table 5. A detailed explanation of the calculation process is presented in the supplementary file. Ad-R3 refers to the adhesive used in the R3 propellant.

Table 4. Contact angles of reference liquids for different adhesive systems.

Sample	Diiodomethane/°	*N,N*-Dimethylformamide/°	Ethylene Glycol/°
Ad-R3	50	58	62
Ad-E3B	62	61	68
Ad-E3D	50	55	55
Ad-E3C	57	55	55

Table 5. The interfacial tension and adhesion work between adhesive or GAP-ETPE and RDX.

Sample	Interfacial Tension/(mJ·m^{-2})	Adhesion Work/(mJ·m^{-2})
Ad-R3	10.6	69.67
Ad-E3B	7.86	66.79
Ad-E3D	7.85	72.59
Ad-E3C	5.14	74.08

Table 5 shows that the magnitude of the adhesion work between the adhesive and RDX is, in ascending order, as follows: Ad-E3B, Ad-E3D, then Ad-E3C. This was consistent with the order of adhesion work observed between the three types of GAP-ETPE and RDX (BDO-ETPE/RDX: 62 mJ·m^{-2}; DBM- ETPE/RDX: 73 mJ·m^{-2}; CBA-ETPE/RDX: 76 mJ·m^{-2}). Table 5 also shows that the adhesion work between the Ad-E3B adhesive system and RDX is lower than that between the Ad-R3 adhesive system and RDX. However, after introducing GAP-ETPE with bonding groups, the adhesion work between the resultant adhesive system (Ad-E3C and Ad-E3D) and RDX is higher than that between Ad-R3 and RDX. These results indicate that the bonding groups can improve the interface performance between the adhesive of a modified double-base propellant and RDX. This improvement may be due to two reasons. First, these elastomers may have formed a homogeneous structure with the double-base adhesive system under the plasticizing effect of Bu-NENA [11]. The elastomers can then be dispersed at the molecular level, allowing them to form effective contact surfaces with RDX. Alternatively, elastomers with bonding groups may have formed strong intermolecular forces with RDX [14]. Specifically, induced force could be produced between the ester groups in DBM and the nitro groups in RDX. Similarly, there could also be an induced force between the cyanide group in CBA and the nitro groups in RDX. It is worth noting that the induced force is stronger in the latter bond due to the stronger polarity of the cyanide group [22] and the increased availability of interaction sites in the CBA-ETPE molecule. Consequently, CBA-ETPE with cyanide as the bonding group was found to be the strongest adhesive system for RDX.

3.2. Tensile Properties of Propellants with Different ETPE Contents

The tensile strength and elongation at maximum strength of the propellant are shown in Figures 2 and 3, respectively. Detailed data on the tensile properties of each propellant can be found in Tables S1–S3. The x in "ExB" refers to the GAP-ETPE content in the EMDB propellant.

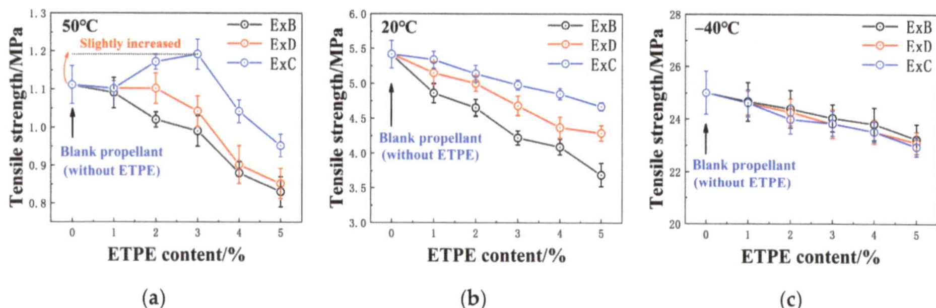

Figure 2. Tensile strength of the EMDB propellant with GAP-ETPE: (**a**) 50 °C; (**b**) 20 °C; (**c**) −40 °C.

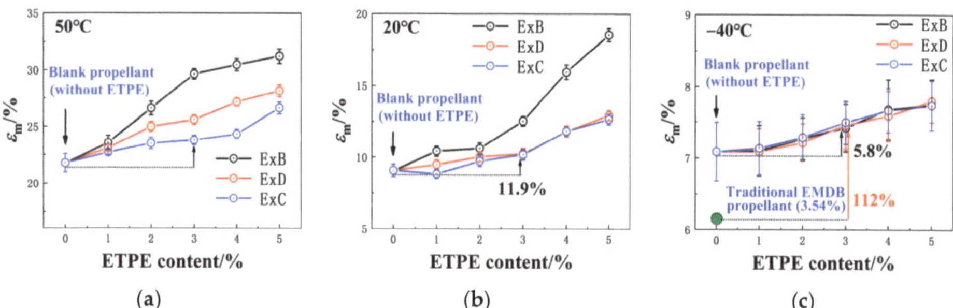

Figure 3. Elongation at maximum strength of EMDB propellant with GAP-ETPE: (**a**) 50 °C; (**b**) 20 °C; (**c**) −40 °C.

Figures 3 and 4 show that the tensile strength of the ExB and ExD propellants at 50 °C decreased with increasing ETPE content, while ExC propellants exhibited an initial increase followed by a decreasing trend. The elongation at maximum strength gradually increased in all three propellants as the GAP-ETPE content increased.

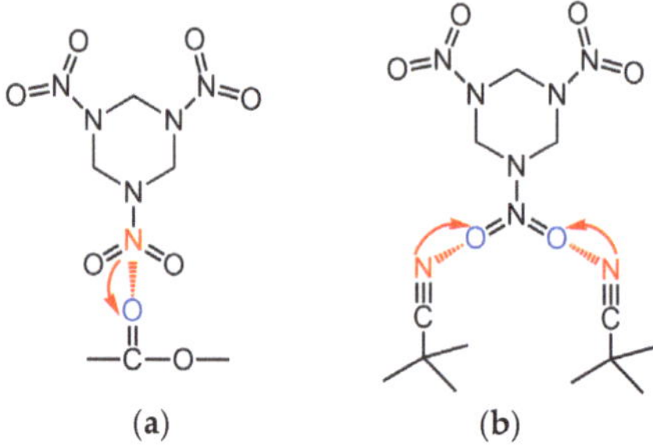

Figure 4. A schematic diagram of the induced forces between RDX and GAP-ETPE: (**a**) the induced force between RDX and DBM-ETPE; (**b**) the induced force between RDX and CBA-ETPE.

Specifically, as the BDO-ETPE content increased, the tensile strength of the ExB propellant at 50 °C decreased while its elongation increased. This is because BDO-ETPE is a typical thermoplastic elastomer with a lower modulus and a greater toughness compared to the adhesive of double-base propellants [23]. The original plasticizing system changes as BDO-ETPE is partially substituted for Bu-NENA/NC in the binder system. Mutual diffusion occurs between NC and ETPE molecules under the plasticizing effect of Bu-NENA [24]. Consequently, a new homogeneous structure is formed comprising three components. Some hydrogen bonds and van der Waals forces between NC molecules are also disrupted by ETPE molecules. With increasing BDO-ETPE content, the reduction in hydrogen bonding within the original binder system is enhanced, leading to promoted molecular mobility [25,26]. Finally, the amount of NC also decreases. Consequently, the modulus of the adhesive declines, supported by the behavior of the tensile modulus shown in Tables S1–S3. In summary, the enrichment of GAP-ETPE in the propellant results in decreased tensile strength and increased elongation.

The trends observed In the ExD series of propellants with Increasing ETPE content were similar to those of the ExB series propellants. However, the tensile strength of ExD propellants was higher while their elongation was lower due to the COOEt group in the DBM molecule, which can induce interactions with the nitro group in RDX (as shown in Figure 4), enhancing the adhesion work between DBM-ETPE and RDX. As indicated in Table 3, the adhesion work between the binder and RDX in the E3D propellant is 72.59 mJ·m^{-2} compared to only 66.79 mJ·m^{-2} in the E3B propellant. This is clear evidence of the greater tensile strength of the ExD propellant compared to the ExB propellant.

As the CBA-ETPE content increases, the tensile strength of the ExC propellant initially increases before decreasing, while the elongation gradually increases. This behavior is due to the CN group in the CBA molecule, which can induce interactions with RDX (as shown in Figure 4). When CN groups were introduced into the binder system, the interactive forces at the binder–RDX interface were enhanced, leading to an increase in the tensile strength of the propellant [25,26]. However, the interfacial area between the adhesive and RDX and the content of the bonding groups in the elastomer was held constant throughout this process; consequently, the improvement in the mechanical properties of the propellant due to CBA-ETPE was non-linear. In addition, as the content of the elastomer increased, the tensile modulus of the binder system decreased (Tables S1–S3), resulting In the reduction of the mechanical properties of the propellant after the CBA-ETPE content exceeded 3% [1]. Since the adhesion work between CBA-ETPE and RDX is greater than that of DBM-ETPE and BDO-ETPE with RDX, the corresponding adhesion work between the Ad-ExC binder and RDX must also be greater than that of the latter two materials with RDX. Therefore, the ExC propellant has the highest tensile strength and lowest elongation compared to ExB and ExD propellants. It is important to consider $\varepsilon_b/\varepsilon_m$, known as the dewetting ratio [1]; a smaller dewetting ratio indicates better interface adhesion, as shown in Tables S1–S3. The ExC propellant has the smallest dewetting ratio, providing further evidence that this category of propellants exhibits the most desirable bonding effect with RDX for CBA-ETPE.

Figures 3 and 4 show that as the temperature decreases, the structural differences of ETPE have less impact on the mechanical properties of the propellant. This is because the detachment of the binder and solid filler occurs prior to the internal fracture of the binder when the propellant is under external load [1,25]. These bonding groups, thus, increase the adhesive work between the binder and RDX, causing the propellant to preferentially undergo an internal tearing of the binder rather than detachment when subjected to external load. The binder system has a larger chain segment movement space at a relatively high temperature, making it prone to slipping between segments, resulting in larger deformations and easier interface detachment with RDX [26,27]. The bonding groups could enhance the interaction between the binder and RDX, preventing detachment in these scenarios. Consequently, these bonding groups have a significant impact on the mechanical properties of the propellant. However, the chain segment movement of the binder is restricted at low temperatures due to the significant intermolecular friction [26,27], making it more susceptible to chain fracture under external load. The weak interface between adhesive and RDX is not the key factor influencing the fracture of the propellant. In other words, the low-temperature tensile performance of the propellant was mainly affected by the content of ETPE rather than the type of ETPE.

3.3. Impact Mechanical Properties

Figure 5 shows the impact strength of EMDB propellants with GAP-ETPE. Detailed data on the impact strength of the propellants can be found in Table S4. Figure 5 also shows that the impact strength of the three types of EMDB propellants gradually increased as the GAP-ETPE increased across all test temperatures. This may be primarily due to the influence of the characteristics of the molecular structure of the GAP-ETPE on the adhesive of the propellant. In the context of molecular structures, GAP-ETPE consists of two parts: a hard and soft segment. The soft segments mainly come from the polyether segments in GAP, which account for 70% of the total segments in the material. The hard segment is mainly composed of amino acid esters obtained by the reaction of isocyanates with chain extenders, accounting for 30% of the total material. The polyether segments are relatively flexible and exhibit good compatibility with NC. When introduced into the EMDB propellant, this material forms a blended system with NC under the plasticizing effect of Bu-NENA [11,28]. In this way, some hydrogen bonding and van der Waals forces between NC molecules are replaced by the van der Waals forces between ETPE and NC molecules, decreasing the resistance of adhesive molecules to movement. At the same time, the relative flexibility of the soft segments in GAP-ETPE increases the ability of the adhesive molecule to change conformations, consequently enhancing the molecular mobility of the adhesive. The improved mobility of the adhesive molecules and the long chains of ETPE make them prone to sliding rather than tearing when subjected to impact loads. Therefore, the impact resistance of these materials was improved.

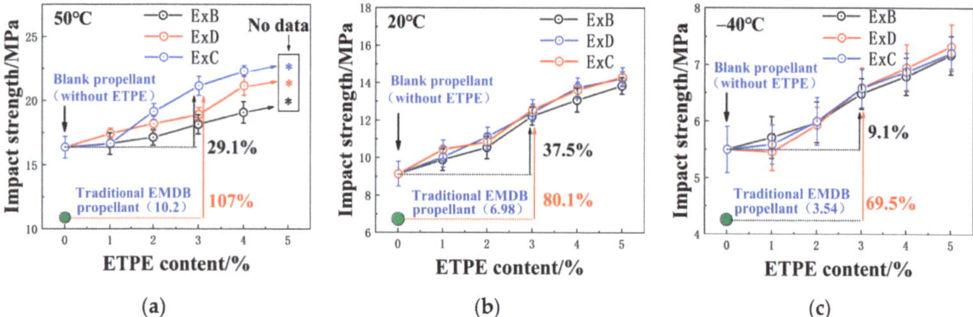

Figure 5. Impact strength of the EMDB propellant with GAP-ETPE: (**a**) 50 °C; (**b**) 20 °C; (**c**) −40 °C. No data refers to samples that were not broken into two pieces.

Figure 5 also shows that the ability of the three elastomers to improve the impact strength of the propellant decreases as temperatures decrease. At high temperatures, the movement space of the binder chain segments is larger and the interaction forces between the segments are weaker. The adhesive molecules have stronger mobility at 50 °C compared to −40 °C. The structural damage suffered during the deformation of the propellant under stress is often caused by the dewetting of the adhesive and RDX. This has been shown using scanning electron microscope (SEM) images of the cross-section of the R3 propellant, which can be found in Figure S2. Notably, the bonding groups of GAP-ETPE can delay the onset of dewetting, increasing the tendency of the propellant adhesive system to undergo structural damage. Consequently, CBA-ETPE, which has the strongest bond with RDX, could maximize the increase in the impact strength of the propellants. Although it is difficult to prove the existence of a delay in the dewetting described above to the direct use of SEM photos, the smaller dewetting ratio observed during the tensile fracture of the ExC propellant at 50 °C indirectly confirms this. As temperatures fall, the reduced intermolecular free volume and internal energy of the material at low temperatures restrict the molecular mobility of adhesive molecules. The increased resistance to molecular motion makes the binder undergo internal tearing more readily. This can be directly proven by the

SEM image of the cross-section of the R3 propellant, as shown in Figure S3. Indeed, it is clear that the interactive forces induced by the bonding groups between the binder and RDX cannot delay the breaking of the adhesive. Consequently, it is difficult for bonding groups to generate the extra increase in the impact strength of the propellant at low temperatures.

3.4. Mechanical Sensitivity

Figure 6 shows the characteristic height (H_{50}) of the EMDB propellant with GAP-ETPE. Detailed data on the sensitivity of the propellants can be found in Table S5. Figure 6 shows that the H_{50} of the EMDB propellant increases as the GAP-ETPE content increases. This suggests that GAP-ETPE has a positive effect on the insensitive properties of the EMDB propellant. It may be attributed to the insensitive nature and the structural characteristic of the ETPE molecule. Based on the results from the literature and the well-known hotspot theory [29–31], we suggest that the propellants undergo plastic deformation when subjected to mechanical stimuli, potentially forming hotspots that can lead to explosions. During the plastic deformation process, defects and voids may form within the binder system, and the closure of these voids, in a manner similar to adiabatic compression, can generate compression hotspots, potentially leading to initiation. In addition, the RDX may experience friction, shear, and compression between particles during the deformation process, leading to the formation of hotspots that can cause initiation. For EMDB propellants with high solid content, the rigid binder and the large amount of RDX can make the propellant more sensitive. The modulus of the binder system may be decreased due to GAP-ETPE, enhancing the toughness of the propellant (as evidenced by tensile test results). While this can promote the continuous plastic deformation of the binder, reducing the generation of internal defects, the reduction of its modulus means that the binder system between particles is more likely to deform and absorb external energy, reducing the rigid friction and compression between particles, allowing it to act as a buffer for the solid fillers. In other words, GAP-ETPE could partially replace RDX in its role in absorbing the energy from external stimuli, reducing the stimulation received by RDX. Together, these factors can reduce the mechanical sensitivity of the propellant.

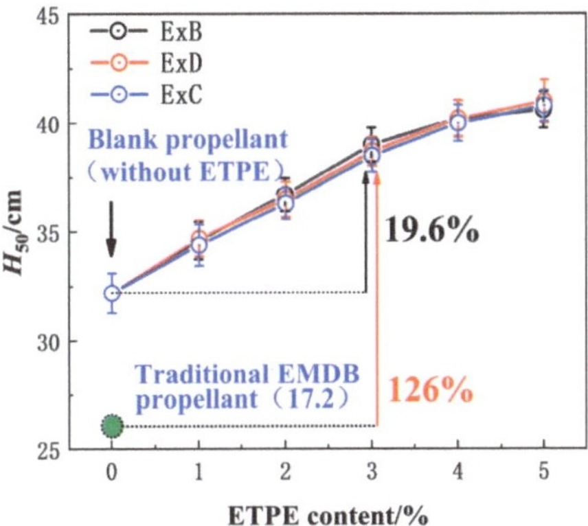

Figure 6. H_{50} of the EMDB propellant with GAP-ETPE.

Figure 6 also shows that the different elastomers had little effect on the H_{50} of the propellant. It indicates that the bonding groups in the ETPE molecule do not influence the sensitivity of the EMDB propellant. This is consistent with the aforementioned mechanical properties.

It should be noted that the R3 propellant was obtained by replacing NG in the traditional composition of the EMDB propellant with Bu-NENA, which is not sensitive to frictional stimuli. Consequently, the friction sensitivities of the three series of propellants under the influence of GAP-ETPE were all 0%. Compared to traditional EMDB propellants, the friction sensitivity of EMDB propellants with GAP-ETPE was reduced from 46% to 0%, while their H_{50} increased by 138% from 17.2 cm to a maximum of 41.0 cm. The friction sensitivity remained at 0% while the impact sensitivity of the propellant was further reduced when comparing the EMDB propellants to the R3 propellant. Generally, the use of GAP-ETPE results in low-sensitivity EMDB propellants.

4. Conclusions

In this study, a novel method of improving the interface performance between adhesives and the RDX through the incorporation of GAP-ETPE with bonding groups was proposed. The effects of molecular structure and content of ETPE on the performance of the EMDB propellants were studied. The main results and conclusions are as follows:

(1) Under the induced force between the bonding groups and the nitro groups of RDX, the interfacial compatibility between the adhesive system of the propellant and the RDX was improved, and its adhesion work increased. The cyanide group, which possesses a stronger degree of polarity, can generate stronger induced forces between the ETPE and the RDX compared to the ester groups. Consequently, the adhesive of the ExC propellant exhibited greater adhesion work with RDX compared to other propellants.

(2) All three types of ETPE improved the low-temperature elongation and impact resistance of the EMDB propellant, but BDO-ETPE and DBM-ETPE also reduced the high-temperature tensile strength of the propellant. Due to its ability to induce stronger forces, CBA-ETPE improved the mechanical properties of the propellant at low and high temperatures simultaneously. The optimal mechanical properties of the propellant were achieved at a 3% CBA-ETPE content (i.e., E3C). The elongation and impact strength of the propellant at $-40\ ^{\circ}$C reached 7.49% and 6.58 MPa, respectively, while its tensile strength and impact strength at 50 $^{\circ}$C reached 1.19 MPa and 21.1 MPa, respectively.

(3) All three types of ETPE improved the safety of the EMDB propellant. As the ETPE content increased, the sensitivity of the EMDB propellant gradually decreased. The optimal formulation for the E3C propellant had a characteristic fall height of 39.0 cm, which is 126% higher than that of the traditional EMDB propellant. In addition, the friction sensitivity of the material was reduced to 0%.

In summary, structural modification of the propellant adhesive using GAP-ETPE with bonding groups could improve the interface performance between the binder and RDX while also enhancing the mechanical properties of the propellant, including its tensile and impact strength. In other words, the novel method proposed in this work represents a simple and effective means of improving the mechanical properties of EMDB propellants, which could provide guidance for studies aiming to optimize the performance of EMDB propellants.

Supplementary Materials: The following supporting information can be downloaded at: https://www.mdpi.com/article/10.3390/polym16060792/s1, Figure S1: The synthesis process diagram of three kinds GAP-ETPE: (a) BDO-ETPE; (b) DBM-ETPE; (c) CBA-ETPE; Table S1: The tensile properties of EMDB propellants at 50 $^{\circ}$C with GAP-ETPE; Table S2: The tensile properties of EMDB propellants at 20 $^{\circ}$C with GAP-ETPE; Table S3: The tensile properties of EMDB propellants at $-40\ ^{\circ}$C with GAP-ETPE; Table S4: The impact strength of EMDB propellants with GAP-ETPE; Table S5: The mechanical sensitivity of EMDB propellants with GAP-ETPE; Figure S2: SEM image of cross section R3 propellant at 50 $^{\circ}$C; Figure S3: SEM image of cross section R3 propellant at $-40\ ^{\circ}$C.

Author Contributions: Y.L. and B.Z. contributed to the conception of the study; S.S., H.L. and Y.W. performed the experiment; W.D. and S.S. performed the data analyses and wrote the manuscript; all authors helped perform the analysis with constructive discussions and reviewed the manuscript. All authors have read and agreed to the published version of the manuscript.

Funding: This research is supported by Shanxi Province Science Foundation for Youths (No. 202203021222073).

Institutional Review Board Statement: The research did not require ethical approval.

Data Availability Statement: Data are contained within the article.

Conflicts of Interest: The authors declare no conflicts of interest.

References

1. Tan, H.M. *The Chemistry and Technology of Solid Rocket Propellant*; Beijing Institute of Technology Press: Beijing, China, 2015; pp. 114–420.
2. Yuan, Q.; Shao, Z.; Zhang, Y. Application of cellulose glycerol ether nitrate in double base propellants. *J. Solid Rocket Technol.* **2012**, *35*, 83–87.
3. Jiao, Q.; Li, J.; Ren, H.; Wang, L.; Zhang, L. Effect of RDX Particle Size on Properties of CMDB Propellant. *Chin. J. Energ. Mater.* **2007**, *15*, 220–223.
4. Wu, C.; Liu, Y.; Hu, S.; Lu, Y.; Guo, C.; Li, H.; Qu, H.; Fu, X.; Li, H. Correlation between microstructural evolution and mechanical properties of CMDB propellant during uniaxial tension. *Prop. Explos. Pyrotech.* **2023**, *48*, e202300117. [CrossRef]
5. Elbasuney, S.; Fahd, A.; Mostafa, H.E.; Mostafa, S.F.; Sadek, R. Chemical stability, thermal behavior, and shelf life assessment of extruded modified double-base propellants. *Def. Technol.* **2018**, *14*, 70–76. [CrossRef]
6. Fleming, J.; Rousseau, W.; Zebregs, M.; van Driel, C. Solventless extruded double-base (EDB) propellant charges—A review of the properties, technology, and applications. *Int. J. Energ. Mater. Chem. Propul.* **2022**, *21*, 13–46. [CrossRef]
7. Ahmed, M.; Petr, S.; Robert, M.; Jan, Z. Ballistic testing and thermal behavior of cast double-base propellant containing Bu-NENA. *Chin. J. Explos. Propellants* **2017**, *40*, 23–28.
8. An, C.; Li, F.; Wang, J.; Guo, X. Surface coating of nitroamine explosives and its effects on the performance of composite modified double-base propellants. *J. Propul. Power* **2012**, *28*, 444–448. [CrossRef]
9. Jia, X.; Cao, Q.; Guo, W.; Li, C.; Shen, J.; Geng, X.; Wang, J.; Hou, C. Synthesis, thermolysis, and solid spherical of RDX/PMMA energetic composite materials. *J. Mater. Sci. Mater. Electron.* **2019**, *30*, 20166–20173. [CrossRef]
10. Chen, L.; Li, Q.; Wang, X.; Zhang, J.; Xu, G.; Cao, X.; He, W. Electrostatic spraying synthesis of energetic RDX@NGEC nanocomposites. *Chem. Eng. J.* **2022**, *431*, 133718. [CrossRef]
11. Sun, S.; Zhao, B.; Cheng, Y.; Luo, Y. Research on Mechanical Properties and Sensitivity of a Novel Modified Double-Base Rocket Propellant Plasticized by Bu-NENA. *Materials* **2022**, *15*, 6374. [CrossRef] [PubMed]
12. Talawar, M.B.; Sivabalan, R.; Mukundan, T.; Muthurajan, H.; Sikder, A.K.; Gandhe, B.R.; Rao, A.S. Environmentally compatible next generation green energetic materials (GEMs). *J. Hazard. Mater.* **2009**, *161*, 589–607. [CrossRef] [PubMed]
13. Chen, M.; Xu, M.; Liu, N.; Mo, H.; Lu, X. Research progress of energetic thermoplastic adhesives. *Explos. Mater.* **2020**, *49*, 1–8.
14. Zhang, Z. *Synthesis and Characterization of Energetic Thermoplastic Polyurethane Elastomers*; Beijing Institute of Technology Press: Beijing, China, 2015.
15. Zhang, J.; Wang, Z.; Sun, S.; Luo, Y. Preparation and Properties of a Novel High-Toughness Solid Propellant Adhesive System Based on Glycidyl Azide Polymer–Energetic Thermoplastic Elastomer/Nitrocellulose/Butyl Nitrate Ethyl Nitramine. *Polymers* **2023**, *15*, 3656. [CrossRef]
16. *GJB 770B-2005*; Test Method of Propellant. National Defense Industry Press: Beijing, China, 2005.
17. Bihari, B.K.; Kumaraswamy, A.; Jain, M.; Rao, N.P.; Murthy, K.P. Effect of pressure on mechanical properties of composite propellant. *Propellants Explos. Pyrot.* **2021**, *46*, 799–805. [CrossRef]
18. Herder, G.; Weterings, F.; de Klerk, W. Mechanical analysis on rocket propellants. *J. Therm. Anal. Calorim.* **2003**, *72*, 921–929. [CrossRef]
19. Faibish, R.S.; Yoshida, W.; Cohen, Y.J. Contact Angle Study on Polymer-Grafted Silicon Wafers. *J. Colloid Interface Sci.* **2002**, *256*, 341–350. [CrossRef]
20. Young, T. An essay on the cohesion of fluids. *Trans. R. Soc. Lond.* **1805**, *95*, 65–87.
21. van Oss, C.J.; Chaudhury, M.K.; Good, R.J. Monopolar surfaces. *Adv. Colloid Interface Sci.* **1987**, *28*, 35–64. [CrossRef]
22. Linder, B.; Kromhout, R.A. Van der Waals induced dipoles. *J. Chem. Phys.* **1986**, *84*, 2753–2760. [CrossRef]
23. Lemos, M.F.; Bohn, M.A. DMA of polyester-based polyurethane elastomers for composite rocket propellants containing different energetic plasticizers. *J. Therm. Anal. Calorim.* **2018**, *131*, 595–600. [CrossRef]
24. Bhowmik, D.; Sadavarte, V.S.; Pande, S.M.; Saraswat, B.S. An energetic binder for the formulation of advanced solid rocket propellants. *Cent. Eur. J. Energ. Mater.* **2015**, *12*, 145–158.
25. Wang, B.; Liao, X.; Wang, Z.; DeLuca, L.T.; Liu, Z.; Fu, Y. Preparation and Properties of a nRDX-based Propellant. *Prop. Explos. Pyrotech.* **2017**, *42*, 649. [CrossRef]
26. Qi, X.F.; Zhang, X.H.; Guo, X.; Zhang, W.; Chen, Z.; Zhang, J.P. Experiment and simulation of NENA on NC plasticizing. *J. Solid Rocket Technol.* **2013**, *36*, 516–520.
27. Shekhar, H. Effect of Temperature on Mechanical Properties of Solid Rocket Propellants. *Def. Sci. J.* **2011**, *61*, 529–533. [CrossRef]
28. Damse, R.S.; Singh, A. Evaluation of Energetic Plasticisers for Solid Gun Propellant. *Def. Sci. J.* **2008**, *58*, 86–93. [CrossRef]
29. Field, J.E.; Swallowe, G.M.; Heavens, S.N. Ignition mechanisms of explosives during mechanical deformation. *Proc. R. Soc. Lond. A Math. Phys. Sci.* **1982**, *382*, 231–244.

30. Walley, S.M.; Field, J.E.; Palmer, S. Impact sensitivity of propellants. *Proc. R. Soc. Lond. Ser. A Math. Phys. Sci.* **1992**, *438*, 571–583.
31. Choudhari, M.K.; Dhar, S.S.; Shrotri, P.G.; Haridwar, S. Effect of high energy materials on sensitivity of composite modified double base CMDB propellant system. *Def. Sci. J.* **1992**, *42*, 253. [CrossRef]

Article

Non-Isothermal Simulation and Safety Analysis of Twin-Screw Extrusion Process for Synthetizing Glycidyl Azide Polymer-Based Energetic Thermoplastic Elastomer

Junming Yuan [1,*], Yan Liu [1], Jinying Wang [1], Yuan Qu [2], Hu Sun [1], Yue Qin [1] and Nan Wang [1]

[1] School of Environment and Safety Engineering, North University of China, Taiyuan 030051, China; 18635820488@163.com (Y.L.); wjywzhy@126.com (J.W.); sunhudyx@163.com (H.S.); cysl080579@163.com (Y.Q.); xiangbei0621@163.com (N.W.)

[2] Jinxi Industrial Group Co., Ltd., Taiyuan 030027, China; 13593198497@163.com

* Correspondence: junmyuan@163.com

Abstract: In order to study the temperature variation and flow characteristics in the twin-screw reactive extrusion process of synthetizing glycidyl azide polymer-based energetic thermoplastic elastomer (GAP-ETPE), a non-isothermal simulation and a safety analysis were carried out. Firstly, based on the synthesis principle of GAP-ETPE, a mechanical sensitivity test, viscosity test and differential scanning calorimetry (DSC) of GAP-ETPE were carried out. Secondly, a three-dimensional physical model of the intermeshing co-rotating conveying element was established by Gambit. A three-dimensional non-isothermal numerical simulation of the conveying and kneading elements was carried out using FLUENT 19.0 software. The temperature, pressure and shear stress field of conveying and kneading elements with different staggered angles were analyzed and compared. The results show that the maximum temperature of the kneading element is always slightly higher than that of the conveying element at the same rotational speed, but the average temperature in the flow channel is always slightly higher than that of the kneading element. The inlet and outlet pressure difference of the kneading elements with a 90° offset angle is the smallest and the safety is the highest. The shear stress in the flow channel of the conveying element is higher than that of the kneading element as a whole, but the shear stress near the outlet of the 90° kneading element is higher than that in the flow channel of the conveying element. Among the kneading elements, the 90° kneading element has the strongest dispersing and mixing ability, followed by the 60° and 45° kneading elements. According to the thermal and physical parameters of the material, the ignition response time is approximately 6 s, which provides a theoretical guide for the safety design of the GAP-ETPE twin-screw extruder.

Keywords: GAP-ETPE; mechanical sensitivity; conveying element; safety; non-isothermal; simulation

Citation: Yuan, J.; Liu, Y.; Wang, J.; Qu, Y.; Sun, H.; Qin, Y.; Wang, N. Non-Isothermal Simulation and Safety Analysis of Twin-Screw Extrusion Process for Synthetizing Glycidyl Azide Polymer-Based Energetic Thermoplastic Elastomer. *Polymers* **2023**, *15*, 3662. https://doi.org/10.3390/polym15183662

Academic Editors: Guangpu Zhang, Zhengmao Ding and Yanan Zhang

Received: 28 July 2023
Revised: 27 August 2023
Accepted: 3 September 2023
Published: 5 September 2023

1. Introduction

Energetic thermoplastic elastomer (ETPE) has the advantages of excellent mechanical properties, high energy, low sensitivity and repeatability, making it an important material base for the development of high-performance explosives. Currently, ETPE has been used in the polyolefin 13 blended explosive and has been applied to the high-energy propellant, increasing its low-temperature impact strength by more than 120% [1]. The application of a high solid content modified the double-base solid propellants and has significantly improved their low-temperature mechanical properties, and ETPE also has good application prospects in combustible cartridges. However, the current synthesis of ETPE uses the intermittent reaction kettle method, which has problems, such as an unstable product quality, poor interbatch repeatability, low production efficiency and low production capacity, making it difficult to meet the urgent demand for ETPE for new high-performance explosives. Therefore, research should be conducted on the reactive extrusion synthesis process of energetic thermoplastic elastomers to improve the quality of ETPE, shorten its production cycle and rapidly promote the development and application of high-performance explosives, which is also an urgent need for the development of weapons and equipment. The development of solid propellant formulations using various energetic and non-energetic binders with a little emphasis on ETPEs was reviewed by M. L. Chan et al. [2]. The review paper explains the synthesis of a wide range of ETPEs, their propellant and explosive formulations along with the thermal studies of the same. A. G. Stern and H. G. Adolph patented the synthesis process of hydrolysable ETPEs based on BAMO, AMMO and poly caprolactone-poly formal polymers [3]. The advantage of this type of ETPE is to recover the ingredients of propellant, explosive or pyrotechnic compositions at the end of their life cycle by hydrolyzing the binder instead of melting.

Glycidyl azide polymer-based energetic thermoplastic elastomer (GAP-ETPE) is a kind of thermoplastic elastomer, which is synthesized by the end capping reaction of GAP with diisocyanate followed by a chain extension with diol. As one of the most-studied compounds in high-energy prepolymers, GAP not only significantly improves the burn rate of propellants but GAP-ETPE-based propellants also have better creep resistance than conventional thermoset propellants [4]. So, GAP-ETPE has attracted much attention from domestic and foreign energy materials workers [5]. The intermittent co-rotating screw extruder is one of the most commonly used types of polymer compounding equipment, and the variation in the screw extruder process parameters plays a crucial role in the quality of the GAP-ETPE products and the safety of the equipment. In addition, the design of an advanced control system for the intermittent reaction process has enabled the successful application of a non-self-compensating dynamic matrix control method to the reactor experimental setup [6].

Reactive extrusion (REX) is a manufacturing technique that combines traditional melt extrusion with chemical reactions, including polymerization, polymer functionalization and depolymerization (chemical recycling) [7,8]. The transfer screw extruder (TSE) is a key piece of equipment in polymer processing, mainly used as a conveyor, mixer or reactor for particles, granules and viscous fluids [9]. The unique geometry of screw extruders allows the handling of very viscous media; thus, they can be used as continuous reactors to conduct polymerization reactions without using any solvents. This is called reactive extrusion polymerization [10]. Sobhani, H. and Verma, S. K. et al. [11,12] combined the virtual domain method with the finite element method to study the flow characteristics of polymer melt in the conveying zone of an intermeshing rotating twin-screw extruder under isothermal conditions. With the development of computational methods, more and more researchers are using numerical simulation to study the non-isothermal flow mechanism in TSEs [13–16]. Ishikawa T. et al. [17] conducted a three-dimensional non-isothermal simulation study of the kneading zone of the co-rotating twin-screw extruder at different speeds and compared the simulation results with the experimental observation to verify the accuracy of the simulation. In addition, Khalifeh et al. [18] investigated the numerical application of non-isothermal flows in a two-channel single-screw extruder using the finite volume method. Therefore, in order to achieve the expected quality throughout the manufacturing process of energetic materials, it is necessary not only to analyze the raw materials and process parameters [19,20] but also to pay special attention to the mechanical properties of these compounds during the extrusion process [21]. The current polymer materials and processing industry heavily rely on single-screw and twin-screw extrusion technology. In order to promote the upgrading of circular economy technology, bridging experimental characterization technology and the predictive ability of modeling and software tools are essential [22].

In order to investigate the effects of operating parameters and screw geometry on the mixing performance during the reactive extrusion process in the TSE, many experimental and numerical simulation studies have been carried out in recent years [23–25]. Ozkan, S. et al. [26] developed a 7.5 mm twin-screw extruder to process high-energy formulations containing nanoparticles. The twin-screw extrusion process has been shown to have several inherent advantages. Rozen [27] used the product distribution of acid-base neutralization and ester hydrolysis to characterize the mixing process of a co-rotating twin-screw extruder. Berzin, F. et al. [28] chose transesterification to optimize and scale up the twin-screw reactive extrusion process. Ke, Z. et al. [29] used the finite element method to analyze the distribution and variation in viscosity during the single-screw extrusion of propellants and studied the threshold and distribution of sensitive parameters, such as pressure and temperature. Zhu, L. and Cegla, M. et al. [30,31] analyzed the polymerization of E-caprolactone in a co-rotating twin-screw extruder using three-dimensional numerical simulation. It is concluded that viscous heat plays an important role in energy generation when the system has high viscosity. Zhang et al. [32] prepared a polypropylene (PP)/wood-fiber (WF) composite foam using the pressure quenching batch method and investigated the effects of the screw configuration, screw speed and silica on the physical and mechanical properties and foaming properties of PP/WF composites. Zong et al. [33] studied the flow pattern, molecular weight distribution and temperature of poly (p-phenylenediamine terephthalate) in three extruders with different elements. Zhang et al. [34] calculated the area draw ratio, instantaneous efficiency and time average efficiency of the mixing disc by studying the distributed mixing power and simulated the local residence time distribution of the TSE using the particle tracking method. Rathod et al. [35] carried out a three-dimensional finite element simulation of a carboxy methyl cellulose aqueous solution in the mixing zone of a twin-screw mixer and studied the influence of the speed, flow rate and blade angle of the mixer on the dispersive mixing of the twin-screw mixer in order to better understand the relationship between the flow, mixing and reaction during polymer preparation. Sayin, Hirata, Hetu and Sun et al. [36–39] carried out three-dimensional numerical simulations in a TSE to investigate the effects of the screw speed, mixing block dislocation, inlet flow rate and stress on the mixing and reaction processes in a TSE. Jong-Sung You et al. [40] characterized the GAP/polyol ETPEs using DSC, dynamic mechanical analysis (DMA) and rheological mechanics. The results show that with an increasing hard segment content both GAP/PTMG ETPEs and GAP/PCL ETPEs exhibit microphase separation transitions.

In the process of mixing energetic materials, the temperature is a key factor affecting the safety of the mixing. If the speed of the impeller is low, the viscosity of the material will be relatively high, the mixing will be inadequate and the expected effect will not be achieved. If the rotating speed of the TSE is high, the viscous heat and shear friction of the material will be greater, the temperature in the channel will be higher, the possibility of combustion and explosion accidents will be greater and the safety will be worse. At the same time, excessive temperatures can volatilize materials and produce flammable gases, which can lead to combustion and/or explosion accidents. In the mixing process, the temperature of the material must be strictly controlled to avoid large temperature fluctuations. Therefore, it is necessary to simulate the temperature field of the flow channel and use the local maximum temperature in the flow channel to characterize the safety of mixing. The main purpose of this paper is to assess the safety of the gauge pressure and critical load pressure obtained from friction susceptibility tests by comparing them with the squeezing pressure and shear stress to which the fluid is subjected in the simulation.

Based on the rheology of the polymer flow, the temperature field, pressure field and shear stress field of the non-Newtonian melt of GAP-ETPE flowing in the twin-screw reactor were analyzed in order to understand the effect of geometric and operational parameters on the reactive extrusion process. In this paper, the GAP-ETPE synthesis process, its performance characterization and a mechanical sensitivity test are carried out. The thermal degradation characteristics of GAP-ETPE are analyzed and the mechanical sensitivity results are obtained. Meanwhile, the three-dimensional physical model of the intermeshing co-rotating conveying element is established using Gambit 2.4 software. Under non-isothermal simulation conditions, the temperature field, pressure field and shear stress field are simulated and analyzed by FLUENT 19.0 software, taking into account factors such as friction, viscous heat generation and synthetic reaction heat. The material flow and temperature changes of GAP-ETPE in the process of conveying and kneading element reactive extrusion are studied using the three-dimensional numerical simulation method; the differences and relationships of the pressure, temperature and shear stress are analyzed; the effects of the screw speed and kneading element pitch angle on the reactive extrusion are discussed; and the ignition response time is calculated according to the thermo-physical parameters of the materials and the convergence of the calculation results. This study provides theoretical guidance for the safety design of the GAP-ETPE twin-screw extruder.

2. Materials and Methods

2.1. Synthesis and Characterization of GAP-ETPE

2.1.1. Synthesis Reaction of GAP-ETPE

GAP-ETPE, Xi'an 204 Institute, Industrial Grade, Shaanxi, China. The synthesis process is shown in Scheme 1. The isocyanate is stored in tank A and the dehydrated polyols, chain extenders and other additives are fully mixed and stored in tank B. The materials are accurately measured and mixed in proportion by the casting machine and then fed into the feed port of the twin-screw reaction extruder. After the twin-screw reaction extruder, the strip-shaped samples are extruded and cut into granules via water-cooled drawing. After forming, the grain is post-cured at room temperature or a high temperature (80 °C or less). The prepared specimen is shown in Figure 1.

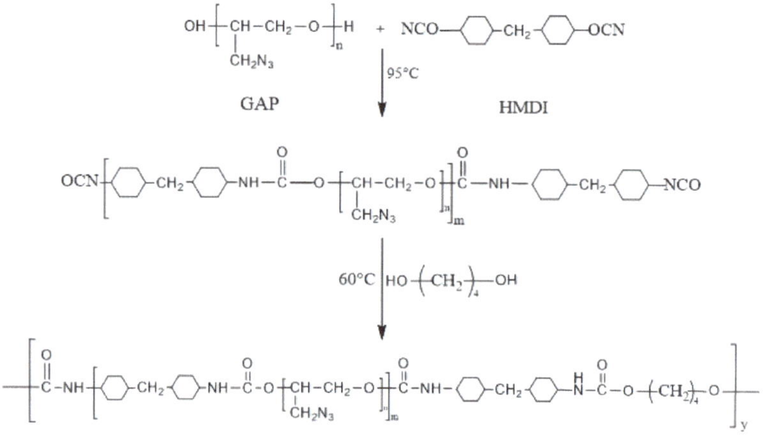

Scheme 1. Synthesis process of GAP ETPE.

Figure 1. Diagram of a sample of granular GAP-ETPE synthesis product.

2.1.2. Performance Characterization and Sensitivity Testing of GAP-ETPE

HTC-1 DSC: Beijing Hengjiu Experimental Equipment Co., Ltd., Beijing, China. The start temperature is set at 30 °C, the end temperature is set at 380 °C, the heating rate is 5 °C/min, the DTA resolution is 0.01 μV, and the sample size is 5–6 mg, with a measurement accuracy of ±2%.

Viscosity test: R/S CPS rheometer, Shanghai Beide Digital Technology Co., Ltd., Shanghai, China. The explosive was loaded into a sealed bottle and dispersed in an ultrasonic constant temperature oil bath at 120 °C for 2 h to form a suspension. The rheological properties of the suspension were then tested using the constant rotation measuring block in the R/S CPS rheometer. The temperature was 120 °C, the shear rate was $2\ s^{-1}$, the number of test points was 60, and the test time was 60 s, rotational viscometer accuracy of ±0.1% and repeatability of ±0.2%.

Mechanical sensitivity: The shock sensitivity is tested according to GJB772A-1997 [41] with a dose of 30 mg, a drop height of 25 cm and a drop weight of 10 kg. The friction sensitivity is tested in accordance with GJB772A-1997 with a dose of 20 mg, a gauge pressure of 2.45 MPa and a swing angle of 66°.

2.2. Establishment of Finite Element Model

2.2.1. Finite Element Model

The twin-screw extruder consists of conveying and kneading elements. Before using FLUENT for simulation calculations, the geometry model of the conveying and kneading elements must be created and meshed using Gambit 2.4 software. The modeling of the conveying element is based on the geometry of intermeshing co-rotating twin screws, and the flow channel model uses Boolean subtraction to remove the conveying element. The physical and finite element models of the screw geometry are shown in Figure 2.

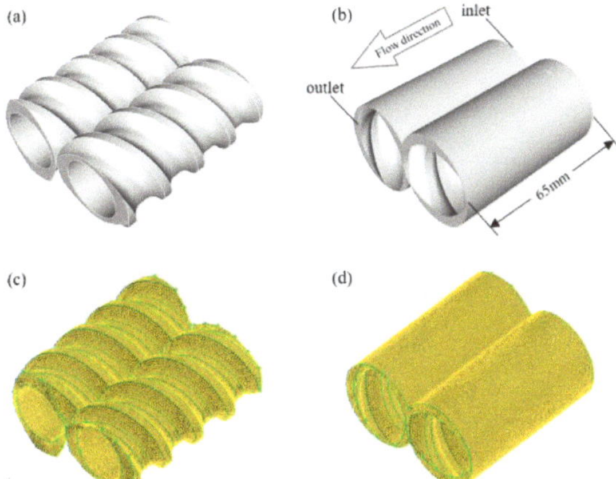

Figure 2. Screw geometry and finite element model: (**a**) screw geometry; (**b**) flow channel geometry; (**c**) finite element model of screw; (**d**) finite element model of flow channel.

In order to compare the temperature changes of the conveying and kneading elements, the kneading elements with different pitch angles and the conveying elements were modeled separately. The screw parameters are given in Table 1. The kneading element is the main part of the mixing efficiency of the TSE and the main part of the temperature change in reactive extrusion. The pitch angle of the kneading element has a great influence on the temperature change. In this section, the typical stagger angles of 45°, 60° and 90° of kneading elements are selected to study effect of changes in the stagger angle on the extrusion temperature of the TSE reaction. Figure 3a,d are the kneading element and channel model with a stagger angle of 45°, i.e., S45; Figure 3b,e are the kneading element and channel model with stagger angle of 60°, i.e., S60; Figure 3c,f are the kneading element and channel model with stagger angle of 90°, i.e., S90. The kneading element parameters are given in Table 2.

Table 1. Twin-screw structure parameters.

Screw Diameter/mm	Screw Clearance/mm	Lead Range/mm	Sleeve Length/mm	Screw Length/mm
26	0.4	32	65	65

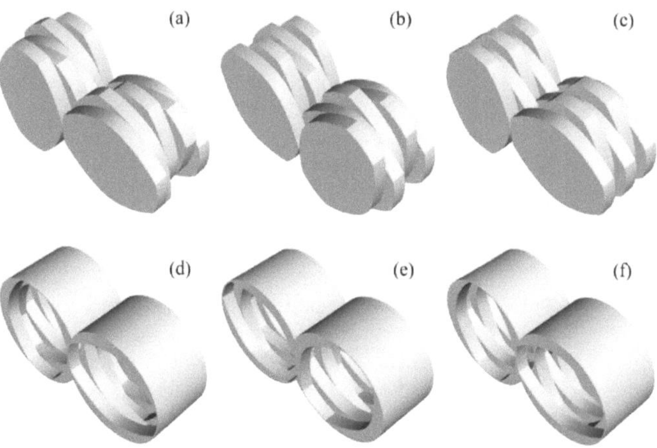

Figure 3. Geometry model of kneading element: (**a**) S45 geometry model; (**b**) S60 geometry model; (**c**) S90 geometry model; (**d**) geometric model of S45 flow channel (**e**) geometric model of S60 flow channel; (**f**) geometric model of S90 flow channel.

Table 2. Structure parameters of kneading element.

Name	Disc Thickness (Number/mm)	Stagger Angle/°	Involute End Circle Diameter/mm	Minimum Tooth Width/mm	Maximum Actual Tooth Width/mm
S90	5/4	90	13	1.49	1.63
S60	5/4	60	13	1.49	1.63
S45	5/4	45	13	1.49	1.63

2.2.2. Mathematical Model

The extrusion movement of the fluid in the flow channel is very complex. Based on the flow characteristics of the polymer in the mixing section of the conveying element and the theory of fluid mechanics, the environment is assumed. The mathematical model of the flow channel is assumed to satisfy the following conditions: (1) the fluid is a power-law fluid; (2) the flow is laminar; (3) the fluid is an incompressible fluid; (4) the flow field is non-isothermal; (5) the viscous heat release of the material in the mixing process is considered; (6) the influence of inertia and gravity is neglected; (7) the non-slip condition is satisfied on the surface of the screw and the inner surface of the barrel.

Based on the above assumptions, the flow chart of the numerical simulation is established as shown in Figure 4 below. However, the assumptions are relatively idealized, so the accuracy of the simulation results has a certain error compared to the experimental values.

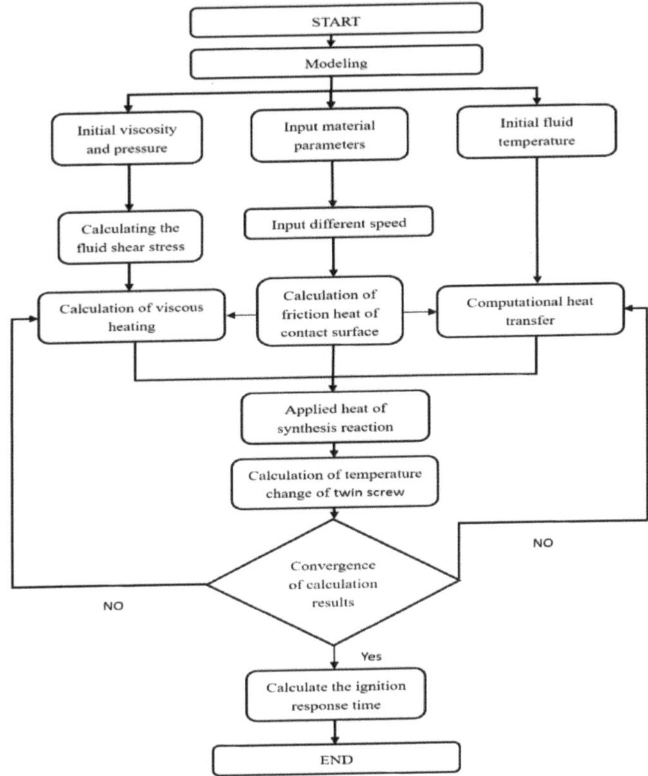

Figure 4. Flow chart of simulation calculation.

The continuity equation is expressed as follows:

$$\frac{u_r}{r} + \frac{\partial(u_r)}{\partial r} + \frac{\partial(u_\theta)}{\partial \theta} + \frac{\partial(u_z)}{\partial z} = 0 \tag{1}$$

The momentum equations are expressed as follows:

$$\tau_{xy} = 2\mu\frac{\partial u_x}{\partial x} + \lambda\nabla\cdot\vec{u} \tag{2}$$

$$\tau_{yy} = 2\mu\frac{\partial u_y}{\partial y} + \lambda\nabla\cdot\vec{u} \tag{3}$$

$$\tau_{zz} = 2\mu\frac{\partial u_z}{\partial z} + \lambda\nabla\cdot\vec{u} \tag{4}$$

$$\tau_{xy} = \tau_{yx} = \mu\left(\frac{\partial u_x}{\partial y} + \frac{\partial u_y}{\partial x}\right) \tag{5}$$

$$\tau_{xz} = \tau_{zx} = \mu\left(\frac{\partial u_x}{\partial z} + \frac{\partial u_z}{\partial x}\right) \tag{6}$$

$$\tau_{yz} = \tau_{zy} = \mu\left(\frac{\partial u_y}{\partial z} + \frac{\partial u_z}{\partial y}\right) \tag{7}$$

Where μ is the dynamic viscosity, Pa·s; λ is the second viscosity, Pa·s, usually $-2/3$; p is the pressure on the fluid microelement, Pa; and τ_{ij} is the stress tensor component, Pa.

Energy conservation refers to the rate of increase of the energy in the microelement body, which is equal to the net heat flux entering the microelement body plus the work performed on the microelement body by the mass force and the surface force. The expression is as follows:

$$\rho c_p\left(\frac{\partial T}{\partial x}+\frac{\partial T}{\partial y}+\frac{\partial T}{\partial z}\right)=k\left(\frac{\partial^2 T}{\partial x}+\frac{\partial^2 T}{\partial y}+\frac{\partial^2 T}{\partial z}\right)+\frac{1}{2}\eta\dot{\gamma} \tag{8}$$

Where p is the pressure, Pa; ρ is the density, g/cm^3; k is the thermal conductivity, W/(m·K); c_p is the specific heat capacity, J/(kg·K); η is the viscosity, Pa·s; and $\dot{\gamma}$ is the shear rate, s^{-1}.

In order to reduce the reaction time and speed up the reaction rate, the amount of catalyst added in the twin-screw reaction process is higher than the conventional amount. Since most of GAP-ETPE is in a viscous state throughout the twin-screw reaction process, the viscosity of GAP-ETPE is taken as the mean value. Figure 5a shows the relationship between the $ln\ \eta^*$ of GAP-ETPE and time, with a median of 6 selected for $ln\ \eta^*$, as shown by the red dot in Figure 5a. Therefore, in this simulation, the material property value of GAP-ETPE is selected as follows: $c_p = 2223$ J/(kg·K), $\rho = 1183$ kg/m^3, $\eta = 400$ Pa·s, as shown by the red dot in Figure 5b.

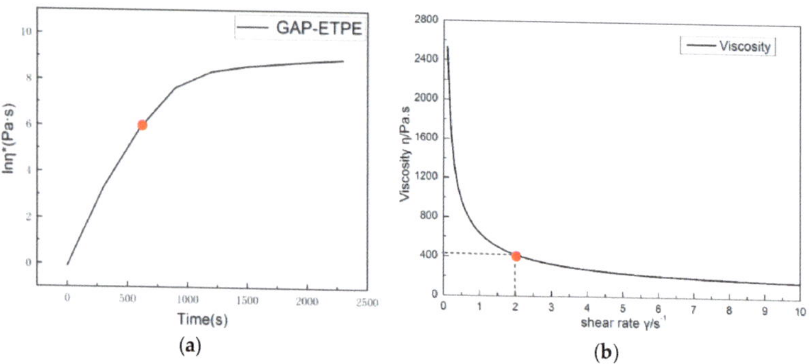

Figure 5. (a) Relationship between $ln\ \eta^*$ of GAP-ETPE versus time; (b) power-law fluid model.

The viscosity of non-Newtonian fluids varies with the shear rate and shear stress, as shown in Figure 5b, so the ratio of shear stress to shear rate at a certain point on the flow curve is used to represent the viscosity at a certain value. This viscosity is called apparent viscosity, which is determined by η symbolic representation. The typical power-law model description commonly used for non-Newtonian fluids is shown in the following formula:

$$\eta = k\dot{\gamma}^{n-1} \tag{9}$$

where k is the consistency index, also known as the consistency coefficient. The value of k is a measure of viscosity, but it is not equal to the viscosity value, and the higher the viscosity, the higher the value of k: $k = 144.43$ Pa·s. $\dot{\gamma}$ is the shear rate, s^{-1}. The power-law exponent n, an index of the flow behavior or the non-Newtonian nature of the material, is a temperature-dependent parameter, and the greater the deviation of n from 1, the more non-Newtonian the material: $n = 0.132$. So, the final power-law fluid constitutive equation is $\eta = 144.43\dot{\gamma}^{-0.868}$.

2.2.3. Mesh Generation

Gambit 2.4 software was used to construct the three-dimensional model of the twin-screw extruder, in order to reduce the workload of simulation calculations. The inside of the twin screw was hollowed out, and the conveying element and the flow channel adopt a mixed mesh of tetrahedron and hexahedron. The interval size of the inner wall is 2, and the interval size of the thread contact surface of the conveying element and the flow channel is 1. To capture the flow characteristics in the gap, two layers of boundary layer are set on the outer wall of the channel, the gap is 0.4 mm and the growth factor is 1.2. The finite element model of the flow channel consists of 199,806 elements and 47,433 nodes. The finite element model of a screw consists of 276,775 elements and 65,706 nodes. The kneading elements with different staggered angles use the same mesh generation method and mesh density.

2.2.4. Boundary Conditions

The rotation of the impeller causes the instability to grow in the fluid. The time selection is transient, and the dynamic mesh is set for the region of the impeller. To account for the heat generated by wall friction during rotation, a roughness constant of 0.5 and a roughness height of 10 mm are set on the contact surface, and the initial temperature is set to 100 °C. The Simple algorithm is selected in FLUENT.

2.2.5. Prepare UDF and Apply Reaction Heat

The thermal decomposition temperature of ETPE is set at 453 K. According to reference [42], in stiu FT-IR was used to monitor the changes in the peak absorption of -NCO during the synthesis reaction of GAP-ETPE to calculate the kinetics of the reaction. On the basis of the Arrhenius Law Equation (10) and Eyring Law (11), the thermodynamic parameters of ETPE, such as the reaction activation energy (E_a), activation enthalpy (ΔH) and activation entropy(ΔS), can be determined, and they are shown in Table 3. These parameters were programed into UDF and imported into FLUENT to prepare for the calculation of the ETPE ignition response time.

$$\ln K = \ln A - \frac{E_a}{RT} \tag{10}$$

$$\ln \frac{K}{T} = -\frac{\Delta H}{RT} + \frac{\Delta S}{R} + \ln \frac{R}{Nh} \tag{11}$$

where K is the reaction rate constant at temperature T, min^{-1}; E_a is reaction activation energy, generally considered as a temperature-independent constant, J/mol or kJ/mol; T is the absolute temperature, K; ΔH is the activation enthalpy, also known as the heat of the reaction (when ΔH is negative, it is an exothermic reaction; when ΔH is positive, it is an endothermic reaction), kJ/mol; ΔS is activation entropy, J/(mol·K); R is the molar gas constant, 8.314 J/(mol·K); N is the Avogadro's constant ($N = 6.02 \times 10^{23}$); and h is the Planck's constant ($h = 6.62 \times 10^{-34}$ J·s).

Table 3. Thermodynamic parameters of ETPE.

E_a(kJ/mol)	A	ΔH(kJ/mol)	ΔS(J/mol^{-1}·K^{-1})
108.996	1.7×10^{10}	1044	−59.397

3. Results and Discussion

3.1. Analysis of Test Results

3.1.1. Thermal Behavior of GAP-ETPE

The DSC and TG-DTG curves of pure GAP-ETPE at the heating rate of 5 °C/min are shown in Figure 6. As shown in Figure 6a, without any melting endothermic processes, the thermal decomposition process of GAP-ETPE shows a single exothermic peak at around 249 °C. With a molar reaction activation enthalpy of 1044 KJ/mol, the main decomposition process starts at around 180 °C and stops at around 280 °C. As shown in Figure 6b, the GAP-ETPE sample undergoes a main exothermic reaction during the second stage, which is confirmed by the DSC curves. The first decomposition stage starts at 40 °C and ends at 75 °C, with a mass loss of about 12%. The second decomposition stage starts at 220 °C and ends at 275 °C with a mass loss of approximately 46.4%. The mass loss in the second decomposition stage is closed to the total mass of the nitrogen content and can therefore be attributed to the release of N_2 from the azide group via the cleavage of the azide bond. The third stage starts at about 290 °C and stops at 370 °C, which does not represent a significant exothermic effect. And the third stage experiences a nearly 30% mass loss due to the fracture of carbon backbones within the soft and hard segment molecular chains.

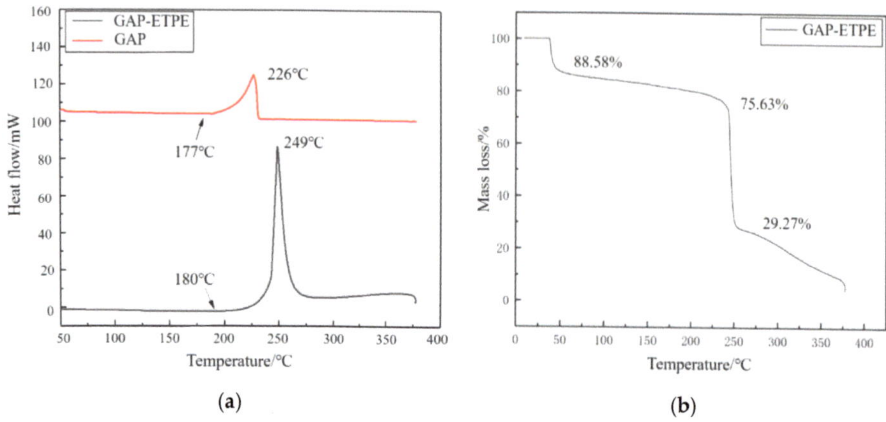

Figure 6. Thermal behavior of GAP-ETPE: (**a**) DSC curve of GAP and GAP-ETPE; (**b**) TG-DTG curve of GAP-ETPE.

3.1.2. Mechanical Sensitivity Analysis

The results of the impact and friction sensitivity tests for GAP-ETPE are shown in Table 4. From Table 4, it can be seen that the impact sensitivity value of GAP-ETPE is 0% and the friction sensitivity value is 15%. Based on the self-designed pendulum impact force testing system [43], the pendulum impact force was tested under the conditions of a 66° pendulum angle and 2.45 MPa overpressure, and the graph of the pendulum impact force versus time was obtained. From Figure 7, it can be seen that the peak value of the pendulum impact force reaches 1540 N. Since the diameter of the sliding column is 0.01 m, the action area is 7.85×10^{-5} m^2. Finally, the ratio of the pendulum impact force to the effective area is 19.6 MPa, which is the shear stress of the GAP-ETPE sample, and the corresponding explosion rate of the friction sensitivity is 15%.

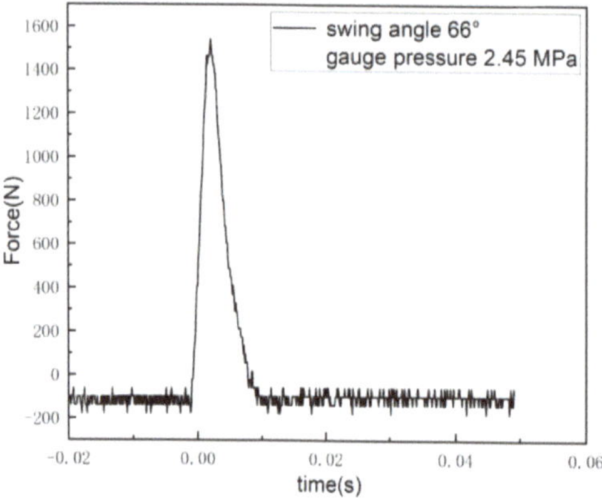

Figure 7. The impact force curve of pendulum of GAP-ETPE friction sensitivity.

Table 4. Test results of impact sensitivity and friction sensitivity of GAP-ETPE propellant.

Sample	Impact Sensitivity/%	Friction Sensitivity/%
GAP-ETPE	0	15

3.2. Calculation Results and Safety Analysis

3.2.1. Discussion and Safety Analysis of Temperature

In order to understand the details of mixing and temperature variations in an intermeshing co-rotating twin screw, a number of monitoring planes have been selected. Figure 8 shows a sketch of an example plane for an intermeshing co-rotating twin screw. The coordinate origin is at the bottom of the screw and fifteen planes are selected as sample planes, i.e., plane 1 (Z = 0 mm)–plane 15 (Z = 65 mm) from the Z axis, to monitor the changes in temperature, pressure and shear stress at each plane.

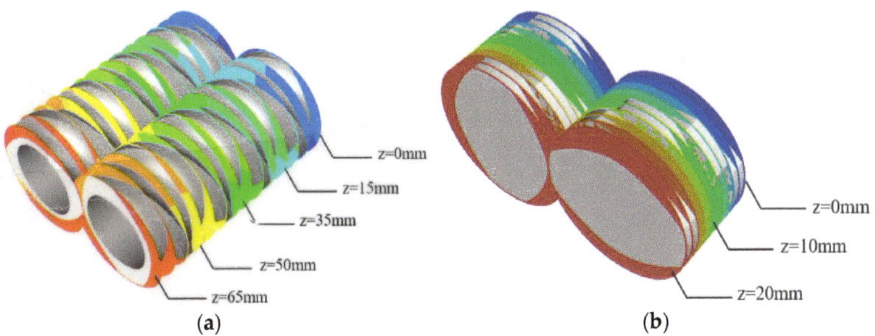

Figure 8. Monitoring plane sketch: (**a**) conveying element monitoring plane sketch; (**b**) kneading element monitoring plane sketch.

In order to reflect the fluid temperature changes in the twin-screw sleeve intuitively, the temperature distribution cloud diagram of the flow channel is processed by slicing, and the profile of ETPE temperature changes is obtained, as shown in Figure 9. It is found that the temperature rise points are mainly concentrated in the reactant mesh area, ETPE was obtained at the distance of 16 mm from Z axis, and it was found that the temperature rise point was concentrated on the contact surface between the screw and the flow channel and decreased gradually from inside to outside. With the increase in screw speed, the temperature in the flow channel is also changing under the conditions of viscous heat and friction. At 40 rpm, the temperature rises from 373 K to 382 K; at 50 rpm it rises to 387 K, at 60 rpm it rises to 394 K and at 70 rpm it rises to 401 K, as shown in Figure 10a. The twin-screw extruder is designed to be approximately 1710 mm and is evenly divided into eight barrels and one screw head, resulting in a length of approximately 190 mm for each segment. The simulated length of the conveying element is 65 mm in this paper, and it covers the three main reaction areas in the middle of the twin-screw extruder, mainly referring to the synthesis reaction area of conveying element barrels 4, 5 and 6. The research group conducted four temperature monitoring experiments at different speeds of 70, 60, 50 and 40 rpm, with two temperature monitoring points on each barrel. The monitoring temperatures of the conveying element for speeds of 70, 60, 50 and 40 rpm are approximately 132 °C (405 K), 124 °C (397 K), 119 °C (392 K) and 110 °C (383 K), respectively. The simulated predicted temperatures of the conveying element for the corresponding four rotational speeds are 128 °C (401 K), 121 °C (394 K), 114 °C (387 K) and 109 °C (382 K), respectively. From the comparison and analysis of predicted temperature results and actual monitoring temperature data, it can be seen that the temperature simulation results of the conveying element are in good agreement with the actual temperature monitoring results and have reliable accuracy.

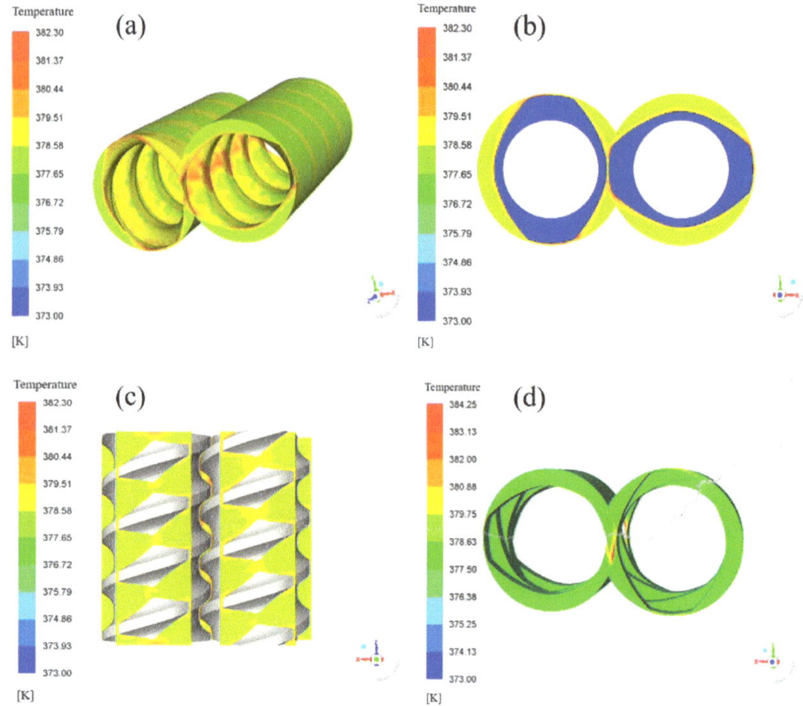

Figure 9. Temperature distribution at 40 rpm: (**a**) conveying element flow channel temperature distribution; (**b**) cross-section of temperature distribution of conveying element flow channel 16 mm from z axis; (**c**) profile of conveying element flow channel temperature distribution; (**d**) flow channel temperature distribution of kneading element.

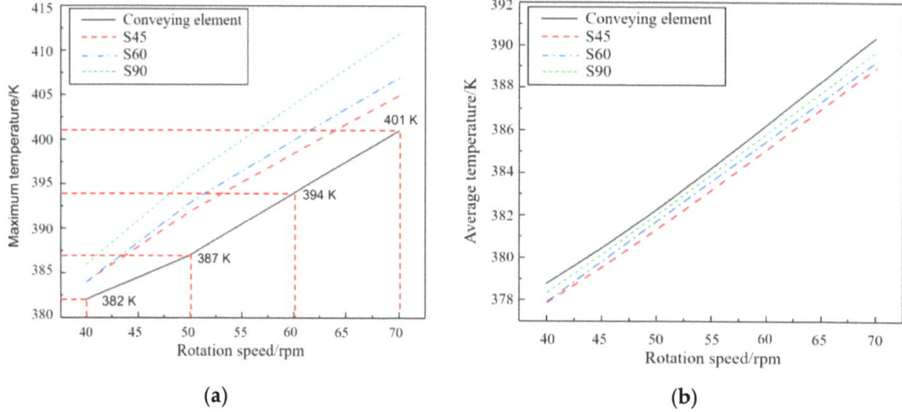

Figure 10. Effect of different rotation speeds on temperature: (**a**) maximum temperature variation curve in flow channel; (**b**) average temperature change curve in flow channel.

In addition, as kneading elements S45, S60 and S90 have different intersection angles, the reactants in this mixing stage undergo strong shear, friction and compression effects, resulting in a significant conversion of mechanical energy into thermal energy. At this point, it is also easy to generate hot spots, as shown in the highest temperature in Figure 10a.

At this stage, the highest temperature of the S90 kneading element, also known as the hot spot, is 4~12 K higher than the highest temperature of the conveying element, with a maximum temperature range of 386–413 K (113–140 °C). Compared with the initial thermal decomposition temperature of GAP-ETPE at 453 K (180 °C), the twin-screw extrusion process is still within the safe range at this time.

It can be seen that at low speed the temperature in the flow channel changes slowly. As the speed increases, the temperature clearly changes. The heat generated by mechanical friction between the conveying element and the flow channel is no longer the main source of energy change, and with the increase in shear stress, the viscous dissipation heat generated by GAP-ETPE gradually takes a dominant position. Figure 9 shows that the temperature rise of the conveying element and the kneading element is mainly concentrated in the meshing area, which is due to the shear and friction between the conveying element and the kneading element and the flow channel during the rotation process, resulting in the accumulation of energy and a higher temperature than other areas. Figure 10a shows that at the same rotational speed the maximum temperature of the kneading element is always slightly higher than that of the conveying element, and the order of the temperatures from highest to lowest is 90° kneading element, 60° kneading element, 45° kneading element and conveying element, but the average temperature in the flow channel is slightly higher than that of the kneading conveying element, as shown in Figure 10b. This is due to the fact that the shear, friction and extrusion are the main sources of heat in the engaging unit, with higher local hot spots, whereas the conveying unit is mainly reactive and exothermic. Therefore, the average temperature in the flow channel is slightly higher than that in the transport element.

3.2.2. Discussion and Safety Analysis of Pressure

From the pressure cloud diagram of the conveying element in Figure 11, it can be seen that the pressure gradually increases along the extrusion direction and there is a higher pressure value at the outlet, the fluid is squeezed by the screw in the flow channel, and the pressure gradually increases. The inlet pressure is the lowest and the outlet pressure is the highest, reaching 25.19 bar, which is 8 times higher than that of the kneading element. When the pressure difference is large, the material flows faster and stays in the flow channel for a shorter time. Therefore, it can be concluded that the residence time of the material in the flow channel of the kneading element is longer than that in the flow channel of the conveying element, so that the material in the flow channel of the kneading element can be mixed more completely.

Figure 12a shows that the higher the speed of the conveying element, the higher the pressure in the flow channel and the greater the pressure difference between the inlet and outlet. The pressure at the inlet of the conveying element is negative, and the further away from the coordinate axis, the greater the pressure. Near the outlet, the pressure value drops sharply, and the higher the speed, the greater the drop, but it is still in accordance with the law that the higher the speed, the higher the pressure. Figure 12b–d shows that as the rotational speed of the kneading element increases, the maximum pressure in the flow channel is greater and the pressure difference between the inlet and outlet is greater. However, the pressure of the different types of kneading elements is also different, with S45 being the largest, S60 the second-largest and S90 the smallest. In S90, the material flow speed is slow, the residence time is long and the mixing is more sufficient. It can be seen from Figure 12d that the maximum fluid pressure of GAP-ETPE is 0.08 MPa during the whole process of twin-screw transport and engagement, which is much lower than the overpressure of 2.45 MPa of the friction sensitive device. This safety factor of GAP-ETPE in the process of twin-screw extrusion is relatively high.

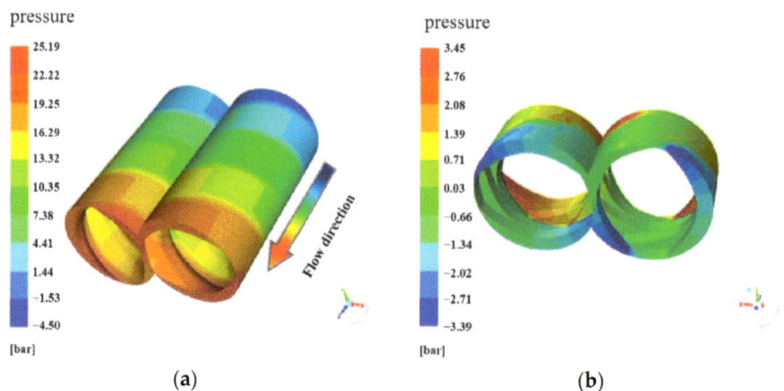

Figure 11. Pressure distribution at 40 rpm: (**a**) conveying element pressure distribution; (**b**) kneading element pressure distribution.

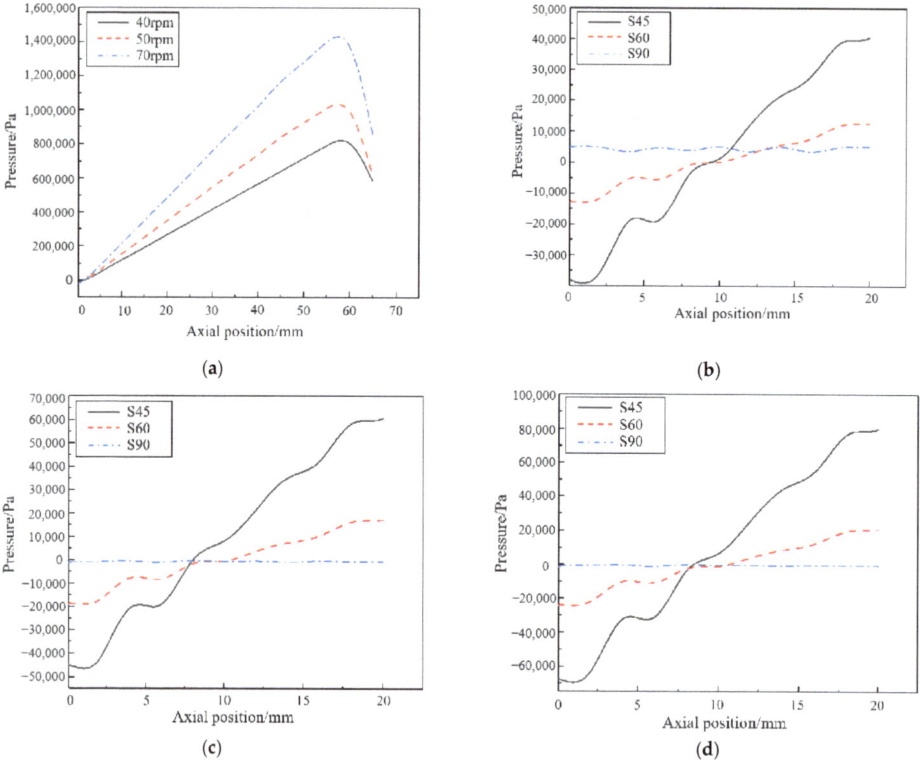

Figure 12. Effect of rotating speed on pressure: (**a**) effect of different rotating speeds on conveying element pressure; (**b**) effect of 40 rpm on kneading element pressure; (**c**) effect of 50 rpm on kneading element pressure; (**d**) effect of 70 rpm on kneading element pressure.

3.2.3. Discussion and Safety Analysis of Shear Stress and Velocity

From the shear stress cloud diagram in Figure 13 of the conveying element and the kneading element, the shear stress is greatest in the meshing area, there is a higher shear stress near the top of the screw, and there is a lower shear stress area near the screw groove. From the top of the screw to the root of the screw groove, the material is subjected to shear stresses from high to low. Figure 14 shows that the shear stress of the conveying element is higher than that of the kneading element when the rotational speed is fixed, the shear stress becomes larger and larger as the rotational speed increases, and the dispersing and mixing ability is stronger. Among the kneading elements, the 90° kneading element has the strongest dispersing and mixing ability, the 60° kneading element takes the second place and the 45° kneading element is the worst. As the rotational speed increases, the shear stress in the flow channel of the impeller increases and has a greater variation at the inlet and outlet. It can be seen from Figure 14d that the maximum shear stress of GAP-ETPE is about 0.005 MPa during the whole process of twin-screw transport and engagement, which is much smaller than the impact force of 19.6 MPa of the friction pendulum under the gauge pressure of 2.5 MPa. Therefore, GAP-ETPE has a high degree of safety in the twin-screw extrusion process.

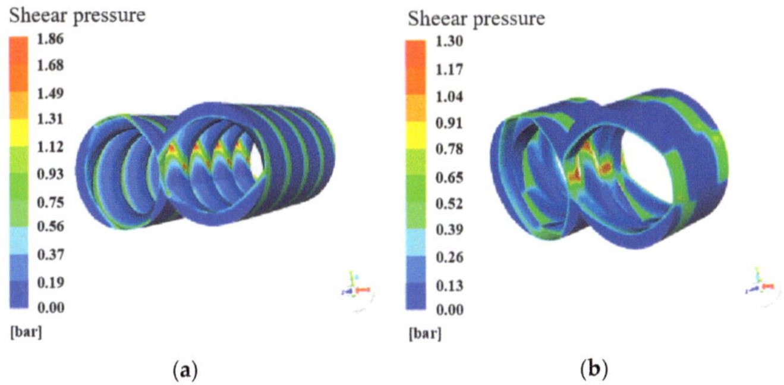

Figure 13. Shear stress distribution at 40 rpm: (**a**) conveying element shear stress distribution; (**b**) kneading element shear stress distribution.

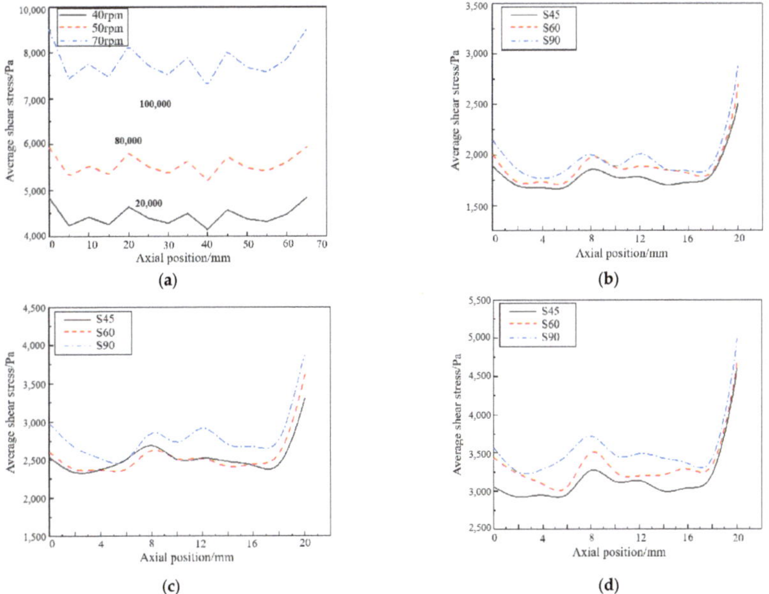

Figure 14. Effect of rotating speed on shear stress: (**a**) effect of different rotating speeds on conveying element shear stress; (**b**) effect of 40 rpm on shear stress of kneading element; (**c**) effect of 50 rpm on shear stress of kneading element; (**d**) effect of 70 rpm on shear stress of kneading element.

Figure 15a shows that the speed of the conveying element is greatest in the meshing area, and then the speed of the fluid contact area with the screw edge is also greater, which is conducive to the complete mixing of the materials, and the fluid area of the kneading element also follows this rule. Figure 15b shows the velocity distribution of the mixing element itself, which increases from the center to either side. The greater the distance, the greater the velocity.

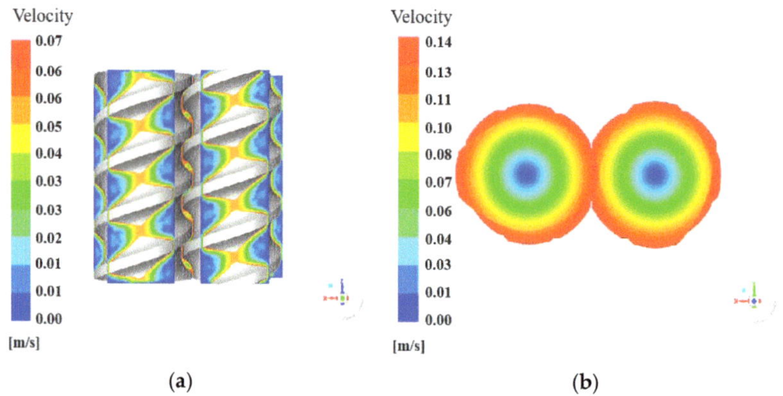

(a)   (b)

Figure 15. Velocity distribution at 40 rpm: (**a**) conveying element velocity distribution profile; (**b**) kneading element velocity distribution.

3.2.4. Discussion and Safety Analysis of Response Time

The thermal decomposition temperature of GAP-ETPE is 453 K (180 °C). The thermophysical parameters of GAP-ETPE in Table 3 are compiled and imported into UDF. When the initial temperature is 453 K, the temperature is unchanged within 1 s, and the temperature increases continuously from 1 s to 5.9 s without any obvious change, but the temperature increases sharply at 5.91 s in Figure 16. It can be seen that ETPE is subject to thermal decomposition. Therefore, to ensure the normal operation of the GAP-ETPE twin-screw extruder, the temperature in the TSE should be lower than 453 K while considering the material mixing uniformity; when the temperature in the TSE reaches 453 K, the mixing equipment should be cooled within 5.9 s to ensure the safety of the equipment. This provides theoretical guidance for the safety design of the GAP-ETPE twin-screw extruder.

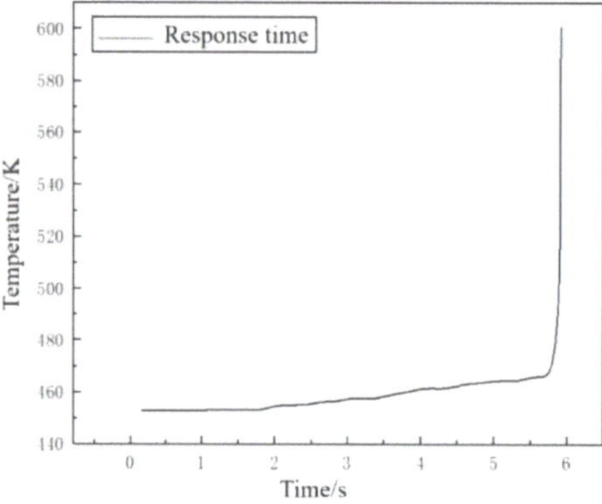

Figure 16. Ignition response time chart.

4. Conclusions

In this paper, a three-dimensional numerical simulation method is used to study the material flow and temperature change of GAP-ETPE during the reactive extrusion of conveying and kneading elements. The following conclusions are drawn:

(1) The heating of the polymer GAP-ETPE comes from the heat conduction in the flow channel, the shear friction between the screw and the flow channel and the heat dissipation caused by the viscosity consumption of the polymer GAP-ETPE. At low speeds, the heat conduction and shear friction in the flow channel are the main factors. As the speed increases, the viscosity consumption and heat dissipation caused by GAP-ETPE gradually become dominant.

(2) At the same speed, the maximum temperature of the kneading element is always slightly higher than that of the conveying element. The order of temperature is 90° kneading element > 60° kneading element > 45° kneading unit > conveying unit, but the average temperature in the flow channel is always slightly higher than the average temperature of the kneading unit. The shear stress of the screw is higher than that of the kneader. As the speed increases, the shear stress increases, resulting in a greater dispersing and mixing ability.

(3) The higher the speed of the conveying and kneading elements, the greater the maximum pressure in the flow channel and the greater the pressure difference between the inlet and outlet. The pressure of the conveying element is higher than that of the kneading element, but the pressure of the kneading element varies with different staggered angles. S45 has the largest pressure change, S60 has the second-largest and S90 has the smallest pressure change. Among them, the pressure difference between the inlet and outlet of the S90 kneading element is the smallest, the residence time is longer and the mixing is more thorough.

(4) The thermal decomposition temperature of GAP-ETPE is 453 K (180 °C). When the initial temperature is 453 K, the temperature is basically unchanged within 1 s, and the temperature rises continuously from 1 s to 5.9 s without any obvious change, but the temperature rises sharply at 5.91 s. In addition, the internal temperature rises of GAP-ETPE synthesized by a twin-screw extruder with a diameter of 26 mm are relatively small under the process conditions of 40–70 rpm, which is much lower than the thermal decomposition temperature of GAPE-ETPE at 180 °C. Therefore, the production process for synthesizing GAP-ETPE through twin-screw reactive extrusion is generally safe.

Author Contributions: Conceptualization, J.Y.; methodology, J.Y.; validation, Y.Q. (Yue Qin), J.W. and H.S.; formal analysis, Y.Q. (Yuan Qu); data curation, J.Y.; writing—original draft preparation, J.Y. and Y.Q. (Yuan Qu); writing—review and editing, J.Y. and Y.L.; visualization, N.W., H.S. and Y.L. All authors have read and agreed to the published version of the manuscript.

Funding: This work was supported by the Robust Munitions Center, CAEP (No. RMC2014B03).

Institutional Review Board Statement: Not applicable to studies that do not involve humans or animals.

Data Availability Statement: The data presented in this study are openly available.

Conflicts of Interest: The authors declare no conflict of interest.

Nomenclature

p	Pressure
τ_{ij}	Stress tensor component
t	Time
ρ	Density
k	Thermal conductivity
C_p	Specific heat capacity
η	Viscosity
$\dot{\gamma}$	Shear rate

K	Reaction rate constant at temperature
E_a	Reaction activation energy
ΔH	Decomposition reaction exothermic
R	Molar gas constant
ΔS	Activation entropy
μ	Dynamic viscosity
λ	Second viscosity
f_x	Unit mass forces in X direction
f_y	Unit mass forces in Y direction
f_z	Unit mass forces in Z direction
u_x	Velocity component in X direction
u_y	Velocity component in Y direction
u_z	Velocity component in Z direction

References

1. Yu, X.H.; Ding, T.; Yu, T.B. Application of energetic binder in PBX explosive under aqueous suspension coating granulation. *Ordnance Ind. Autom.* **2020**, *39*, 82–84.
2. Chan, M.L.; Reed, R.; Ciaramitaro, D.A. Advances in Solid Propellant Formulations. In *Progress in Astronautics and Aeronautics, 185 Solid Propellant Chemistry, Combustion, and Motor Interior Ballistics*; Brill, T.B., Rey, W.Z., Eds.; American Institute of Aeronautics and Astronautics, Inc.: Reston, VA, USA, 2000; pp. 185–206.
3. Stern, A.G.; Adolph, H.G. Hydrolysable Energetic Thermoplastic Elastomers and Methods of Preparation Thereof. U.S. Patent 6395859B1, 28 May 2002.
4. Liu, S.; Zhang, X.-M.; Zhang, J.; Zheng, M.-Z.; Liu, W.-H.; Luo, Y. Creep properties of GAP-ETPE-based energetic solid propellants. *J. Fire Explos.* **2022**, *45*, 877–883. [CrossRef]
5. Tzoganakis, C. Reactive extrusion of polymers: A review. *Adv. Polym. Technol. J. Polym. Process. Inst.* **1989**, *9*, 321–330. [CrossRef]
6. Teng, S. Design and implementation of advanced control system for batch reactor. *Dalian Univ. Technol.* **2023**, *8*, 19.
7. Zhuang, Y.; Saadatkhah, N.; Morgani, M.S.; Xu, T.; Martin, C.; Patience, G.S.; Ajji, A. Experimental methods in chemical engineering: Reactive extrusion. *Can. J. Chem. Eng.* **2023**, *101*, 59–77. [CrossRef]
8. De Smit, K.; Wieme, T.; Marien, Y.W.; Van Steenberge, P.H.; D'hooge, D.R.; Edeleva, M. Multi-scale reactive extrusion modelling approaches to design polymer synthesis, modification and mechanical recycling. *React. Chem. Eng.* **2022**, *7*, 245–263. [CrossRef]
9. Sakai, T. Screw extrusion technology—Past, present and future. *Polimery* **2013**, *58*, 847–857. [CrossRef]
10. Li, T.T.; Feng, L.F.; Gu, X.P.; Zhang, C.L.; Wang, P.; Hu, G.H. Intensification of polymerization processes by reactive extrusion. *Ind. Eng. Chem. Res.* **2021**, *60*, 2791–2806. [CrossRef]
11. Sobhani, H.; Ghoreishy, M.H.R.; Razavi-Nouri, M.; Anderson, P.D.; Meijer, H.H.E. Modelling of polymer fluid flow and residence time distribution in Twin-screw extruder using fictitious domain method. *Plast Rubber Compos.* **2011**, *40*, 387–396. [CrossRef]
12. Verma, S.K.; Gupta, V.; Mukherjee, S.S.; Gangradey, R.; Srinivasan, R. Development of CFD model for the analysis of a cryogenics twin-screw hydrogen extruder system. *Cryogenics* **2021**, *113*, 103232. [CrossRef]
13. Sardo, L.; Vergnes, B.; Valette, R. Numerical modelling of the non-isothermal flow of a non-newtonian polymer in a co-kneader. *Int. Polym. Process.* **2017**, *32*, 425–433. [CrossRef]
14. Malik, M.; Kalyon, D.M.; Golba, J.C., Jr. Simulation of Co-Rotating Twin-screw Extrusion Process Subject to Pressure-Dependent Wall Slip at Barrel and Screw Surfaces: 3D FEM Analysis for Combinations of Forward- and Reverse-Conveying Screw Elements. *Int. Polym. Process.* **2014**, *29*, 52–63. [CrossRef]
15. Salahudeen, S.A.; AlOthman, O.; Elleithy, R.H.; Al-Zahrani, S.M.; Rahmat, A.R.B. Optimization of Rotor Speed Based on Stretching, Efficiency, and Viscous Heating in Nonintermeshing Internal Batch Mixer: Simulation and Experimental Verification. *J. Appl. Polym. Sci.* **2012**, *127*, 2739–2748. [CrossRef]
16. Salahudeen, S.A.; Elleithy, R.H.; AlOthman, O. Comparative study of internal batch mixer such as cam, banbury and roller: Numerical simulation and experimental verification. *Chem. Eng. Sci.* **2011**, *66*, 2502–2511. [CrossRef]
17. Ishikawa, T.; Kihara, S.I.; Funatsu, K. 3-D numerical simulations of nonisothermal flow in co-rotating conveying element extruders. *Polym. Eng. Sci.* **2000**, *40*, 357–364. [CrossRef]
18. Khalifeh, A.; Clermont, J.R. Characteristic curves based on 3D non-isothermal flow computations in single-screw extruder for non-newtonian fluids. *Int. Polym. Process.* **2007**, *22*, 324–333. [CrossRef]
19. Martinez-Pastor, J.; Franco, P.; Franco-Menchon, J.A. Optimization of extrusion process of double-base propellants from their rheological properties. *Int. J. Mater. Form.* **2019**, *12*, 307–320. [CrossRef]
20. Martinez-Pastor, J.; Franco, P.; Ramirez, F.J.; Lopez-Garcia, P.J. Influence of rheological behaviour on extrusion parameters during non-continuous extrusion of multi-base propellants. *Int. J. Mater. Form.* **2018**, *11*, 87–99. [CrossRef]
21. Kalaycioglu, B.; Dirikolu, M.H.; Çelik, V. An elasto-viscoplastic analysis of direct extrusion of a double base solid propellant. *Adv. Eng. Softw.* **2010**, *41*, 1110–1114. [CrossRef]

22. Edeleva, M.; De Smit, K.; Debrie, S.; Verberckmoes, A.; Marien, Y.W.; D'hooge, D.R. Molecular scale-driven upgrading of extrusion technology for sustainable polymer processing and recycling. *Curr. Opin. Green Sustain. Chem.* **2023**, *43*, 100848. [CrossRef]
23. Fel, E.; Massardier, V.; Mélis, F.; Vergnes, B.; Cassagnau, P. Residence time distribution in a high shear Twin-screw extruder. *Int. Polym. Process.* **2014**, *29*, 71–80. [CrossRef]
24. Connelly, R.K.; Kokini, J.L. The effect of shear thinning and differential viscoelasticity on mixing in a model 2D mixer as determined using FEM with particle tracking. *J. Non-Newton. Fluid Mech.* **2004**, *123*, 1–17. [CrossRef]
25. Fard, A.S.; Hulsen, M.A.; Meijer, H.E.H.; Famili, N.M.H.; Anderson, P.D. Tools to Simulate Distributive Mixing in Twin-Screw Extruders. *Macromol. Theory Simul.* **2012**, *21*, 217–240. [CrossRef]
26. Ozkan, S.; Gevgilili, H.; Kalyon, D.M.; Kowalczyk, J.; Mezger, M. Twin-screw extrusion of nano-alumina–based simulants of energetic formulations involving gel-based binders. *J. Energ. Mater.* **2007**, *25*, 173–201. [CrossRef]
27. Rozen, A.; Bakker, R.A. Effect of Operating Parameters and Screw Geometry on Micromixing in a Co-Rotating Twin-Screw Extruder. *Chem. Eng. Res. Des.* **2001**, *79*, 938–942. [CrossRef]
28. Berzin, F.; David, C.; Vergnes, B. Optimization and Scale-Up of Twin-Screw Reactive Extrusion: The Case of EVA Transesterification. *Int. Polym. Process.* **2020**, *35*, 422–428. [CrossRef]
29. Ke, Z.; Zhongqi, H.; Shupan, Y.; Wanghua, C. Numerical simulation for exploring the effect of viscosity on single-screw extrusion process of propellant. *Procedia Eng.* **2014**, *84*, 933–939. [CrossRef]
30. Zhu, L.; Narh, K.A.; Hyun, K.S. Investigation of mixing mechanisms and energy balance in reactive extrusion using three-dimensional numerical simulation method. *Int. J. Heat Mass Transfer* **2004**, *48*, 3411–3422. [CrossRef]
31. Cegla, M.; Engell, S. Application of Model Predictive Control to the reactive extrusion of e-Caprolactone in a twin-screw extruder. *IFAC-PapersOnLine* **2021**, *54*, 225–230. [CrossRef]
32. Zhang, Z.X.; Gao, C.; Xin, Z.X.; Kim, J.K. Effects of extruder parameters and silica on physico-mechanical and foaming properties of PP/wood-fiber composites. *Compos. Part B* **2012**, *43*, 2047–2057. [CrossRef]
33. Zong, Y.; Tang, H.; Zhao, L. 3-D Numerical Simulations for Polycondensation of Poly(p-phenylene terephthalamide) in Twin-screw Extruder. *Polym. Eng. Sci.* **2017**, *57*, 1252–1261. [CrossRef]
34. Zhang, X.M.; Feng, L.F.; Chen, W.X.; Hu, G.H. Numerical Simulation and Experimental Validation of Mixing Performance of Kneading Discs in a Twin-screw Extruder. *Polym. Eng. Sci.* **2009**, *49*, 1772–1783. [CrossRef]
35. Rathod, M.L.; Kokini, J.L. Effect of mixer geometry and operating conditions on mixing efficiency of a non-Newtonian fluid in a conveying element mixer. *J. Food Eng.* **2013**, *118*, 256–265. [CrossRef]
36. Sayin, R.; El Hagrasy, A.S.; Litster, J.D. Distributive mixing elements: Towards improved granule attributes from a conveying element granulation process. *Chem. Eng. Sci.* **2015**, *125*, 165–175. [CrossRef]
37. Hirata, K.; Ishida, H.; Hiragohri, M. Effectiveness of a Backward Mixing Screw Element for Glass Fiber Dispersion in a Twin-Screw Extruder. *Polym. Eng. Sci.* **2014**, *54*, 2005–2012. [CrossRef]
38. Hétu, J.F.; Ilinca, F. Immersed boundary finite elements for 3D flow simulations in twin-screw extruders. *Comput. Fluids* **2013**, *87*, 2–11. [CrossRef]
39. Sun, D.; Zhu, X.; Gao, M. 3D numerical simulation of reactive extrusion processes for preparing PP/TiO2 nanocomposites in a corotating Twin-screw extruder. *Materials* **2019**, *12*, 671. [CrossRef]
40. You, J.S.; Noh, S.-T. Rheological and thermal properties of glycidyl azide polyol-based energetic thermoplastic polyurethane elastomers. *Polym. Int.* **2013**, *62*, 158–164. [CrossRef]
41. *GJB 772A-1997*; Explosive Test Methods. Ordnance Industry Corporation: Beijing, China, 1997; "601.1 Explosion Probability Method" and "602.1 Explosion Probability Method".
42. Wang, J.C. Kinetic Studies on the Polymerization of Dicyclohexylmethylmethane-4, 4′-Diisocyanate with Polyoxytetramethylene Diol. *Adv. Mater. Res.* **2013**, *791–793*, 32–35. [CrossRef]
43. Yuan, J.; Qin, Y.; Peng, H.; Xia, T.; Liu, J.; Zhao, W.; Liu, Y. Experiment and Numerical Simulation on Friction Ignition Response of HMX-Based Cast PBX Explosive. *Crystals* **2023**, *13*, 671. [CrossRef]

Article

Preparation and Properties of a Novel High-Toughness Solid Propellant Adhesive System Based on Glycidyl Azide Polymer–Energetic Thermoplastic Elastomer/Nitrocellulose/Butyl Nitrate Ethyl Nitramine

Jing Zhang [1], Zhen Wang [1], Shixiong Sun [2,3] and Yunjun Luo [1,*]

[1] School of Materials Science and Technology, Beijing Institute of Technology, Beijing 100081, China
[2] School of Chemistry and Chemical Engineering, North University of China, Taiyuan 030051, China
[3] Dezhou Industrial Technology Research Institute of North University of China, Dezhou 253034, China
*   Correspondence: yjluo@bit.edu.cn

Abstract: Glycidyl azide polymer (GAP)–energetic thermoplastic elastomer (GAP-ETPE) propellants have high development prospects as green solid propellants, but the preparation of GAP-ETPEs with excellent performance is still a challenge. Improving the performance of the adhesive system in a propellant by introducing a plasticizer is an effective approach to increasing the energy and toughness of the propellant. Herein, a novel high-strength solid propellant adhesive system was proposed with GAP-ETPEs as the adhesive skeleton, butyl nitrate ethyl nitramine (Bu-NENA) as the energetic plasticizer, and nitrocellulose (NC) as the reinforcing agent. The effects of the structural factors on its properties were studied. The results showed that the binder system would give the propellant better mechanical and safety properties. The results can provide a reference for the structure design, forming process, and parameter selection of high-performance GAP-based green solid propellants.

Keywords: GAP-ETPE/NC/Bu-NENA blend adhesive; Bu-NENA plasticizer; high toughness

Citation: Zhang, J.; Wang, Z.; Sun, S.; Luo, Y. Preparation and Properties of a Novel High-Toughness Solid Propellant Adhesive System Based on Glycidyl Azide Polymer–Energetic Thermoplastic Elastomer/Nitrocellulose/Butyl Nitrate Ethyl Nitramine. *Polymers* **2023**, *15*, 3656. https://doi.org/10.3390/polym15183656

Academic Editor: Antonio Pizzi

Received: 18 July 2023
Revised: 7 August 2023
Accepted: 8 August 2023
Published: 5 September 2023

1. Introduction

Double-base (DB) propellant, one of the earliest developed propellant varieties, is a homogeneous propellant with nitrocellulose as the adhesive and nitroglycerin as the plasticizer. It has been widely applied in solid rocket motors due to its various advantages, such as smokeless products, adjustable energy, abundant sources of raw materials, and mature technology [1]. However, DB propellants have the drawbacks of low-temperature embrittlement, largely caused by semirigid nitrocellulose (NC) macromolecules, and limited energy due to the limited concentration of oxidizer fragments [2,3]. Al and nitramines such as hexogen (RDX) [4], octogen (HMX) [3,5,6], and hexanitrohexaazaisowurtzitane (HNIW or CL-20) [2,5,7] are incorporated into propellant compositions to achieve higher performance. In this way, a modified double-base (MDB) propellant with increased specific impulse (I_{sp}) was achieved [2], as expected. However, the presence of nitroglycerin (NG) and some high-energy materials in MDB propellants makes these systems extremely sensitive [2,5,8,9]. What is more, the toughness of the propellants worsens.

Because of their high production efficiency and low cost, DB and MDB propellants are important components of weapons and equipment. However, their poor safety performance introduces higher risks to their production, transportation, and use. The resulting safety accidents which occur from time to time are mainly caused by two factors: First, a large amount of NG is used as the energetic plasticizer in this type of propellant. It is extremely sensitive to mechanical stimuli and not conducive to the good safety performance of the propellant [10,11]. Second, nitrocellulose (NC), which is used as the adhesive skeleton in these propellants, has high molecular rigidity and strong intermolecular forces, resulting in a high processing temperature for the propellant, complicating its safe formation. The toughness is also very poor, especially for high-energy propellants with solid particles. Therefore, it is of vital importance to improve safety performance through the innovation of formulation components.

Several meaningful explorations have been carried out to reduce the mechanical sensitivity of DB and MDB propellants and improve the processing technology, including solid-filler coating, the optimization of the adhesive structure, and the introduction of additives [1]. Among them, optimizing the adhesive structure is one of the most effective ways to achieve these aims and is performed through the regulation of plasticizers and adhesives [12]. For plasticizers, butyl nitrate ethyl nitramine (Bu-NENA; the structure is shown in Figure 1), nitriethylene glycol (TEGDN), 1, 2, 4-butanol trinitrate (BTTN), trimethylethane trinitrate (TMETN), and other nitrate plasticizers mixed with NG or used alone can achieve the expected goal [1,2]. From molecular structural analysis, it can be determined that the same nitrate group that comprises Bu-NENA has an affinity with NC. Compared with NG, Bu-NENA, which has fewer nitrate groups and more flexible butyl, more easily enters the NC system to plasticize it. For sensitivity, as is known, the nitrate groups in NG are too close, and the groups easily influence each other, resulting in NG's sensitive nature. Its characteristic drop height H_{50} is usually determined as less than 10 cm. For Bu-NENA, there is only one nitrate in a single molecule. The nitroamine group is much more stable than the nitrate group, and the molecule has a relatively long inert butyl group. Consequently, the mechanical sensitivity is significantly lower than that of NG, and the H_{50} can reach more than 110 cm. Therefore, Bu-NENA has a stronger plasticizing ability than NC and a lower mechanical sensitivity [13,14], which can improve the process performance of the system significantly [15–17]. For adhesives, the structural characteristics of NC can realize a narrow range of adjustable process performance. As a result, energetic polymer adhesives with better mechanical properties are available.

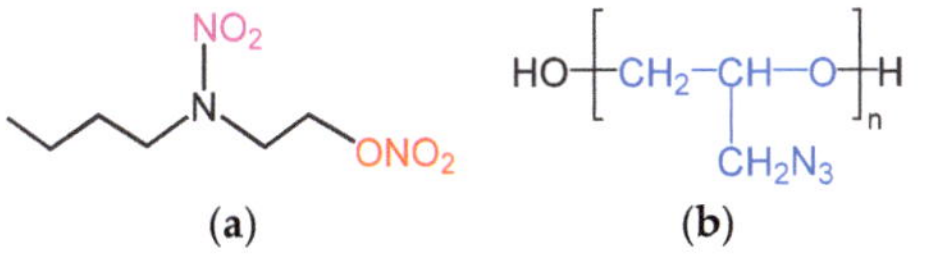

(a)   (b)

Figure 1. Structural formula of (**a**) Bu-NENA and (**b**) GAP molecule.

Energetic thermoplastic elastomer (ETPE) is a two-phase block copolymer consisting of a soft segment and a hard segment, in which the soft segment provides softness and elasticity and the hard segment provides rigidity. These excellent properties make ETPE one of the best choices for an adhesive in propellants [18]. At present, the energetic polymers that have been developed are mainly nitrate polymers: azido-substituted oxysterol derivative polymers; azido cellulose and azide glycidyl ether polymers; and polymers containing nitroamines, nitro groups, and other energetic groups [15–17]. Among them, glycidyl azide polymer (GAP)–energetic thermoplastic elastomers (GAP-ETPEs) have good mechanical properties, low mechanical sensitivity, and a relatively mature preparation process. They are soluble and fusible, increasing their potential as green solid propellant binders. Therefore, developing new solid propellants using GAP-ETPEs as an adhesive remains a current research hotspot [17]. The structural formula of the GAP-ETPE molecule is shown in Figure 2, where R_1 represents the carbon chain of the diisocyanate and R_2 represents the carbon chain of the diol.

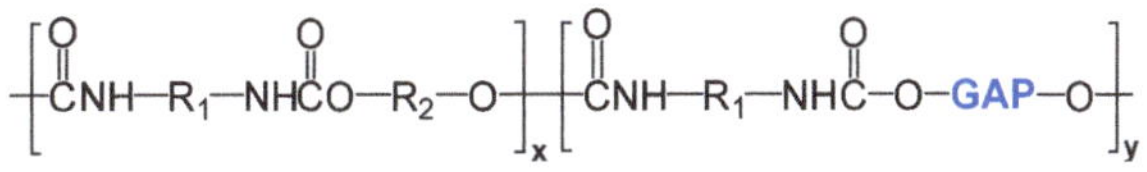

Figure 2. Structural formula of GAP-ETPE molecule.

Based on the advantages of GAP-ETPEs and the urgent need for high-safety propellants in current weapon systems, this paper proposed a novel high-strength solid propellant adhesive system using GAP-ETPEs as the adhesive skeleton, Bu-NENA as the energetic plasticizer, and NC as the reinforcing agent. In our previous study [19], it was found that a GAP-ETPE/NC blend adhesive system experienced a sudden change in the performance of the blend system caused by the change in the NC state when the NC content was 30%. Therefore, in this paper, a GAP-ETPE/NC blend adhesive with 30% NC was selected as the base binder to prepare a GAP-ETPE/NC/Bu-NENA blend adhesive system, and the effects of the structural factors on its properties were studied. The results showed that the binder system was expected to improve the mechanical and safety properties of the propellant. These results can provide a reference for the structure design, forming process, and parameter selection of high-performance GAP-based green solid propellants.

2. Materials and Methods

2.1. Experimental Materials

The GAP-ETPE with 30% hard segment content synthesized in our laboratory has an average molecular weight of about 27,000. Its density was determined as 1.24 g/cm^3, and its enthalpy of formation was calculated as 4828 kJ/mol. The NC content was 12.0%; the average molecular weight was about 80,000 (Shanxi Xingan Chemical Plant (Taiyuan, China)). Bu-NENA, a light yellow oily liquid, had a melting point of −28.0 °C, density of 1.21 g/cm^3, and heat of formation of 249 kJ/mol (Liming Chemical Institute, Luoyang, China).

2.2. Sample Preparation

The synthesis strategy for GAP-ETPEs is shown in Figure 3. A detailed synthesis process can be found in Ref. [18]. Specifically, a stoichiometric amount of GAP was heated and stirred at 90 °C, and then a certain amount of catalyst and HMDI were added. The reaction mixture was stirred and mixed for 2 h at 90 °C. Then, BDO was added to the NCO-terminated GAP prepolymer after the system cooled to 60 °C. The resulting product was cast in a mold, kept for 3–5 min, and then cured at 100 °C for 12 h, after which the GAP-based ETPE with chains extended by BDO was finally obtained.

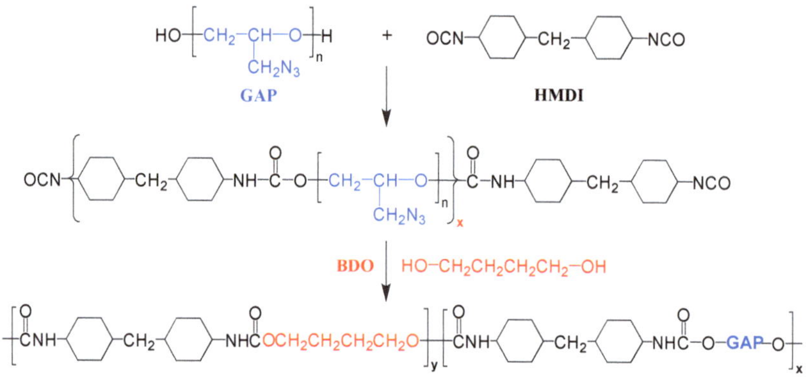

Figure 3. The synthesis process of GAP-ETPE with chains extended by BDO.

At 70 °C, the GAP-ETPE with a certain mass was calendered and plasticized for a certain time on a two-roll mixer, and then the Bu-NENA and NC were mixed, and the mixed Bu-NENA/NC was slowly added to GAP-ETPE in batches with repeated mixing, and then the mixture was calendered for 30 min after the GAP-ETPE was completely added.

The tablets obtained by rolling were placed in the mold, and the mold was placed in the middle position of the platen. At 70 °C, the mold was preheated for 30 min, and the equipment was started to close the mold at a molding pressure of 3–5 MPa for 2 min. The GAP-ETPE/NC/Bu-NENA samples were obtained by opening the mold. The sample with

10% Bu-NENA content was named E-N-N-10, and then E-N-N-15, E-N-N-20, E-N-N-25, and E-N-N-30 (Table 1), with the number indicating the Bu-NENA content. The sample without Bu-NENA was named N30, indicating that the mass percentage of NC was 30 wt%.

Table 1. GAP-ETPE/NC/Bu-NENA blend binders with different Bu-NENA contents.

Samples	GAP-ETPE/NC (7/3)/wt%	Bu-NENA/wt%
N30	100	0
E-N-N-10	90	10
E-N-N-15	85	15
E-N-N-20	80	20
E-N-N-25	75	25
E-N-N-30	70	30

2.3. Experimental Instruments and Test Conditions

2.3.1. Mechanical Property Test

The mechanical properties were tested with an AGS-J electronic universal testing machine (Shimadzu, Kyoto, Japan). Dumbbell splines were prepared according to GB/T528-1998 [20]. The test was performed at room temperature, and the tensile rate was 100 mm/min.

2.3.2. Differential Scanning Calorimetry (DSC)

A DSC1/500/578 differential scanning calorimeter (Mettler-Toledo, Zurich, Switzerland) was used over a temperature range of -100–100 °C at a heating rate of 10 °C/min. The sample mass was 4–5 mg. The sample was heated from room temperature to 100 °C, held for 10 min, then cooled to -100 °C, held for 10 min, and heated to 150 °C at a warming rate of 10 K/min. An N_2 atmosphere was provided at a flow rate of 40 mL/min.

2.3.3. Crosslinking Density Test

A magnetic resonance crosslinking density analyzer (Suzhou Niumai Electronic Technology Co., Ltd., Suzhou, China) was used to test the samples at different temperatures. Each sample was tested 5 times, and the average value was used.

2.3.4. Scanning Electron Microscopy

An S4800 field-emission scanning electron microscope (Hitachi, Tokyo, Japan) was used for scanning electron microscopy (SEM) imaging with a cold field emission electron source, a resolution of 15 kV, and 500× magnification. The scale bar in the images is 100 μm.

2.3.5. Rheological Test

The rheological measurements were performed on a modular advanced rheometer system (Haake, Vreden, Germany) equipped with a 20 mm parallel-plate geometry and a gap width of approximately 1 mm in air. The temperature was controlled with a Haake test chamber controller with an accuracy of ±1 °C. RheoWin Data Manager was used for equipment control, data acquisition, and analysis.

3. Results

3.1. Mechanical Properties of GAP-ETPE/NC/Bu-NENA Blend Adhesives with Different Bu-NENA Contents

The effect of Bu-NENA content on the mechanical properties of GAP-ETPE/NC/Bu-NENA adhesive systems was studied to improve the mechanical properties of the adhesives and expand the scope of mechanical property regulation of adhesive systems by providing a reference for the formulation design of high-performance propellants.

As shown in Figure 4 and Table 2, the maximum tensile strength observed for a GAP-ETPE/NC blend adhesive N30 was 8.7 MPa, which rapidly decreased with increased Bu-NENA. This is mainly because Bu-NENA not only destroyed the physical crosslinking network in the GAP-ETPE/NC system [21–23] but also increased the distance between the GAP-ETPE and the NC, reducing the strength of the intermolecular forces in the GAP-ETPE/NC system.

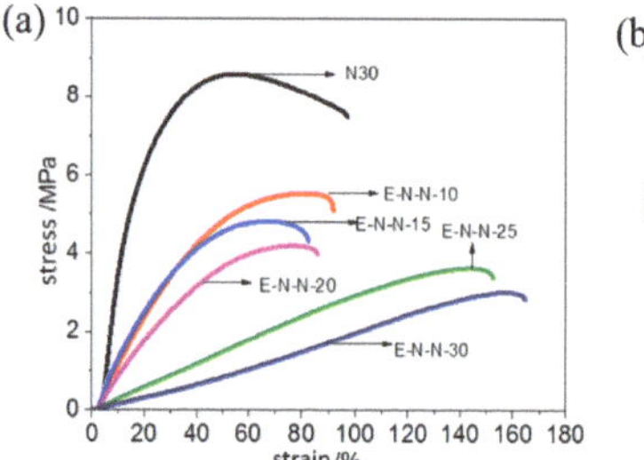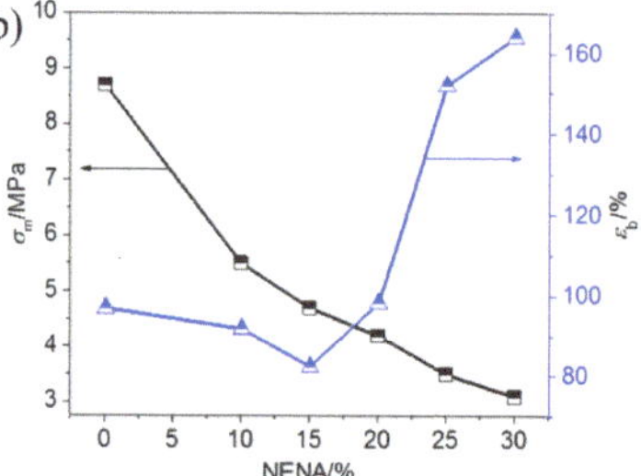

Figure 4. Tensile performance of GAP-ETPE/NC/Bu-NENA systems with different Bu-NENA contents. (**a**) Tensile curve; (**b**) maximum tensile strength and elongation at break. The test temperature was room temperature, and the tensile rate was 100 mm/min.

Table 2. Mechanical property parameters of GAP-ETPE/NC/Bu-NENA systems.

Samples	σ_m/MPa	ε_b/%
N30	8.7	96.8
E-N-N-10	5.5	91.7
E-N-N-15	4.7	82.3
E-N-N-20	4.2	98.2
E-N-N-25	3.5	152.1
E-N-N-30	3.1	164.1

In addition, the elongation at break of the GAP-ETPE/NC/Bu-NENA blends decreased and then increased with increased Bu-NENA, mainly due to the plasticizing effect of Bu-NENA on NC and GAP-ETPE. For low Bu-NENA contents, the plasticization of NC and GAP-ETPE by Bu-NENA was incomplete, which reduced the tensile strength and elongation at break of the blend system. Moreover, a small amount of Bu-NENA could not completely plasticize the NC, and the size of the NC particles dispersed in the GAP-ETPE increased, resulting in stress concentration when the GAP-ETPE/NC/Bu-NENA blending system was subjected to external forces. Thus, the blending adhesive system was easily cracked and destroyed.

The experimental results show that in terms of mechanical properties, the room temperature strength of the adhesive system was adjustable in the range of 3.1–8.7 MPa, and the corresponding elongation was 96.8–164.1%. Generally, the tensile strength of the NC/NG adhesive system at 20 °C was about 10 MPa, and the elongation was about 20% (from DB propellant with NC/NG/C$_2$ 51.5/47.5/1 mass%). Compared with the NC/NG adhesive system, the adhesive system proposed in this paper has a significant advantage in elongation, which is the performance factor critical to such propellants.

3.2. Thermal Transition Behavior of GAP-ETPE/NC/Bu-NENA Blend Adhesives with Different Bu-NENA Contents

DSC was used to investigate the thermal transition behavior of GAP-ETPE/NC/Bu-NENA blend adhesives. Figure 5 and Table 3 show the differential curve and local magnification of the DSC curves and the glass transition temperatures of GAP-ETPE/NC/Bu-NENA blend adhesives with different Bu-NENA contents, respectively. There were two transitions in the DSC curves of all the GAP-ETPE/NC/ Bu-NENA blend adhesives with different Bu-NENA contents before 120 °C, which was the glass transition temperature of the soft and hard segments in GAP-ETPE (Table 3). The T_g^L of the GAP-ETPE/NC blend adhesives was −36.3 °C, and the T_g^L of the GAP-ETPE/NC/Bu-NENA blend adhesives decreased with increased Bu-NENA content. When the Bu-NENA content reached 30%, the T_g^L of the blend adhesives was $-47.5\,°C$.

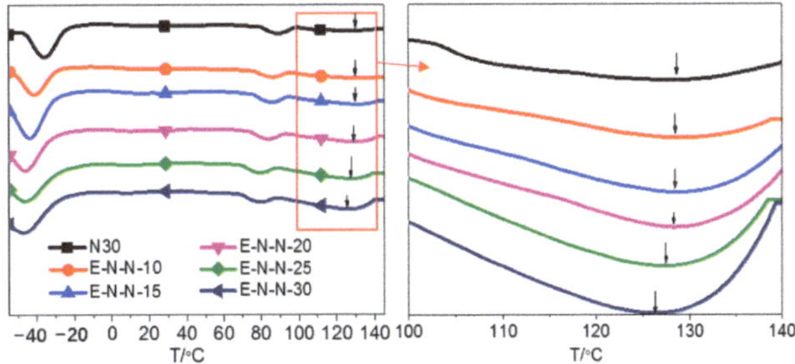

Figure 5. Differential curves of the DSC curves of GAP-ETPE/NC/Bu-NENA blend binders with different Bu-NENA contents. The heating rate is 10 k/min.

Table 3. T_g of GAP-ETPE/NC/Bu-NENA at N30 and different Bu-NENA contents.

Samples	$T_g^L/°C$	$T_g^H/°C$
N30	−36.3	89.4
E-N-N-10	−41.8	85.8
E-N-N-15	−44.5	85.1
E-N-N-20	−46.8	83.4
E-N-N-25	−47.0	80.3
E-N-N-30	−47.5	80.1

The T_g^H of the GAP-ETPE/NC/ Bu-NENA blends also decreased with increased Bu-NENA content due to the reduced intermolecular forces between the GAP-ETPE and NC from the addition of Bu-NENA [23]. This reduction made the molecular movement of GAP-ETPE and NC easier, reducing the T_g of the blend adhesive system. With increased Bu-NENA content, the plasticizing effect of Bu-NENA on GAP-ETPE and NC was enhanced, and the intermolecular forces between the GAP-ETPE and NC were further reduced, decreasing the glass transition temperature of the blend binder system.

Figure 5 shows a small peak (indicated by an arrow) appeared in the GAP-ETPE/NC/ Bu-NENA blend adhesive system above 120 °C. The peak became more prominent and moved to a lower temperature with increased Bu-NENA content. The peak may arise from the T_g of NC, between 170 °C and 180 °C [1]. In the blending system, Bu-NENA could dissolve and plasticize NC [23], and GAP-ETPE also has a plasticizing effect on NC [24], reducing the interaction forces between the NC molecules and between the NC and GAP-ETPE molecules. As a result, the molecular chain segments move easily, decreasing the T_g of NC.

Thus, Bu-NENA can not only plasticize NC but also effectively plasticize GAP-ETPE molecules, reduce the intermolecular forces in the system, increase the free volume of adhesive molecules, and enhance the motility of the adhesive system. This makes it a good candidate as a binder for solid propellants that will improve the toughness of the propellants.

3.3. Physical Crosslinking Density of GAP-ETPE/NC/Bu-NENA Blend Adhesives with Different Bu-NENA Contents

The crosslinking density of a polymer is closely related to the properties of the polymer system. The crosslinking density of the GAP-ETPE/NC/Bu-NENA system could reflect the number of physical crosslinking points in the system. The physical crosslinking points of this system include topological entanglement formed by mutual contact and entanglement among GAP-ETPE molecules, NC molecules, and Bu-NENA molecules, condensation entanglement formed by the interaction between local molecular chains (such as van der Waals forces and hydrogen bonding), and a small number of hard segment aggregation micro-regions in GAP-ETPE [23]. These points and the molecular chain segments of the blend system constitute the physical crosslinking network structure inside the blend adhesive.

Figure 6 shows the crosslinking density of the GAP-ETPE/NC/Bu-NENA blends with different Bu-NENA contents. The physical crosslinking density of N30 (without added Bu-NENA) was 3.44×10^{-4} mol/cm^3. With increased Bu-NENA content, the physical crosslinking density of the blend system continuously decreased. When the Bu-NENA content was 30%, the physical crosslinking density of the blend system decreased to 1.04×10^{-4} mol/cm^3. The main reasons for this decrease were: (1) With the addition of the micro-molecule plasticizer Bu-NENA, the intermolecular force and friction between GAP-ETPE and NC were reduced, and the kinematic ability of the macromolecular chains of the blend system improved [22,23]. The hydrogen bonding and van der Waals forces formed between GAP-ETPE molecules, NC molecules, and GAP-ETPE-NC molecules in the system were destroyed, and the micro-region structure formed by GAP-ETPE hard segment agglomeration deteriorated, which reduced the number of condensed entanglement points in the system [25]. (2) Bu-NENA solvated the topological entanglement points formed between the GAP-ETPE molecules, NC molecules, and GAP-ETPE-NC molecules in the blending system, reducing the number of physical crosslinking points. (3) Bu-NENA increased the intermolecular distances of the blend system and hindered contact between molecular chains and the formation of physical crosslinking points, reducing the crosslinking density of the blend system [26]. In conclusion, the addition of Bu-NENA to the GAP-ETPE/NC system effectively reduced the number of physical crosslinking points and the role of the physical crosslinking network structure [27], affecting the viscosity, mechanical strength, and other properties of the blend system.

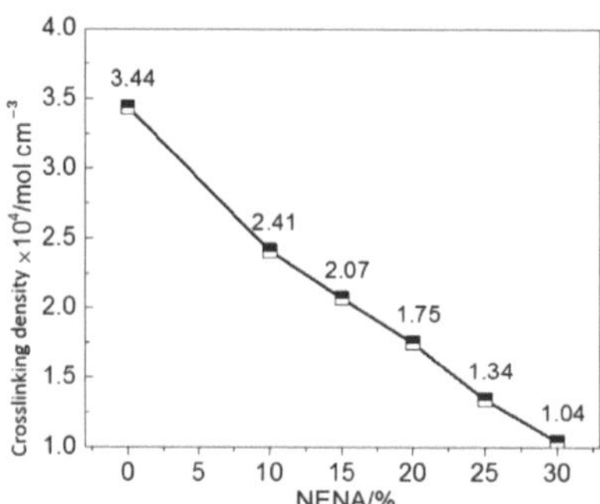

Figure 6. Crosslinking density of GAP-ETPE/NC/Bu-NENA blends with different Bu-NENA contents.

The GAP-ETPE/NC/Bu-NENA adhesive system is complex, and the physical crosslinking points include topological entanglement points formed by mutual contact and intertwining of GAP-ETPE molecules, cohesive entanglement points formed by local molecular chain interactions (such as van der Waals forces and hydrogen bonding), and hard segment aggregation micro-regions with less content in GAP-ETPE. In order to represent the influence of Bu-NENA on the adhesive architecture, we use a schematic diagram and omit the ubiquitous van der Waals forces and entanglements that are difficult to represent simply. We mainly considered the influence of the hydrogen bonds and the hard segment aggregation regions, and the results are shown in Figure 7. In the absence of Bu-NENA, there were more hydrogen bonds between the hard segment aggregation regions of GAP-ETPE and the NC molecules, and there were also some hydrogen bonds between the GAP-ETPE and NC molecules. After the introduction of Bu-NENA, which can plasticize NC and GAP-ETPE, destroy the original hydrogen bonds, and increase the free volume between molecules, the motility of the GAP-ETPE and NC molecular chains improved. From the subsequent SEM imaging, when the content of Bu-NENA reaches a certain level, the NC diffused into the GAP-ETPE, improving the compatibility between the two and removing their phase interface.

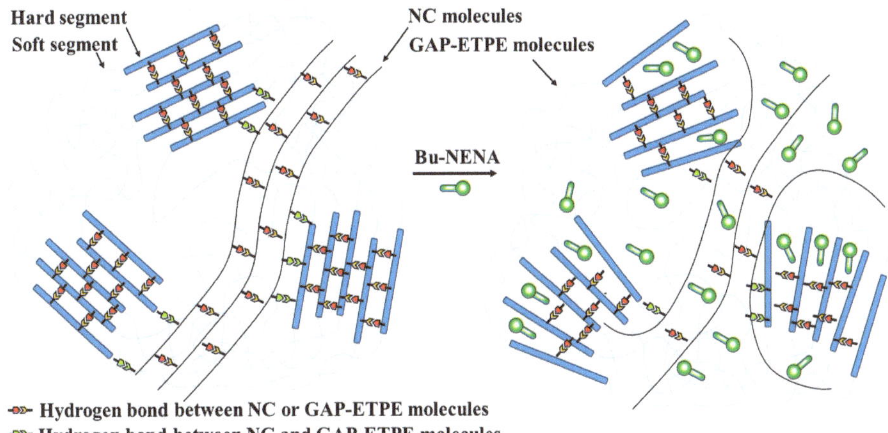

Figure 7. Schematic diagram of the influence of Bu-NENA on adhesive architecture.

3.4. Brittle Surface Morphology of GAP-ETPE/NC/Bu-NENA Blend Adhesives with Different Bu-NENA Contents

Figure 8 shows the SEM images of the brittle surfaces of GAP-ETPE/NC/Bu-NENA blends with different Bu-NENA contents.

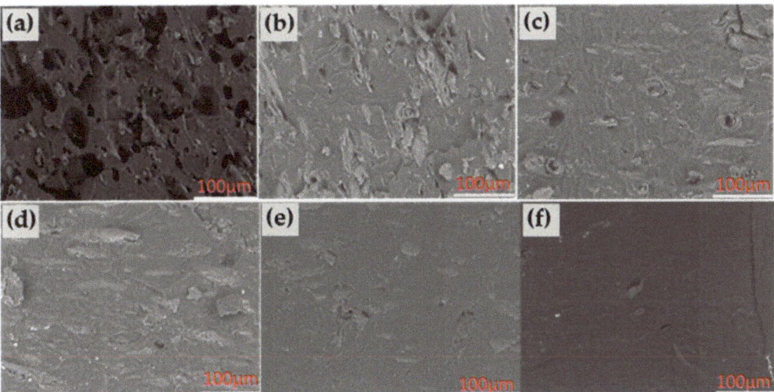

Figure 8. SEM photos of cross-sections of ETPE/NC/Bu-NENA blends with different Bu-NENA contents. (**a**) N30 does not contain Bu-NENA. (**b**) E-N-N-10, (**c**) E-N-N-15, (**d**) E-N-N-20, (**e**) E-N-N-25, and (**f**) E-N-N-30 are ETPE/NC/Bu-NENA blends. The samples were immersed in liquid nitrogen for 5 min and then broken manually, resulting in brittle fracture, before they were observed.

As shown in Figure 8a, the surface of N30 is rough and uneven, with many holes and widely distributed particles. Moreover, the interface between the NC and the GAP-ETPE, caused by their poor compatibility, is obvious. Compared with the GAP-ETPE/NC blend adhesive, the number of holes on the surface of the GAP-ETPE/NC/Bu-NENA blend adhesive is significantly reduced, and the particle accumulation on the surface is larger, as shown in Figure 8b. This is mainly because the Bu-NENA was first mixed with the NC, which increased the plasticity and ductility of the NC. With increasing Bu-NENA content, some NC particles gradually entered the surface of the blend adhesive, and some particles protruded from the surface of the adhesive, resulting in visible fibrous NC, as shown in Figure 8c. With the further increase in Bu-NENA content, the number of particles on the surface of the GAP-ETPE/NC/Bu-NENA blend adhesive decreased, and the surface became flat with almost no raised particles, but an obvious fibrous profile was observed, as shown in Figure 8d. When the content of Bu-NENA increased further (Figure 8e,f), the particles on the surface of the GAP-ETPE/NC/Bu-NENA blend adhesive almost completely disappeared, and the surface became smooth and flat. The NC was completely plasticized by the high Bu-NENA content and changed from the solid state to the sol state, which came into more complete contact with the GAP-ETPE during the blending process. Moreover, Bu-NENA increased the free volume of the molecules in the blending system and the kinematic ability of the GAP-ETPE and NC molecular chains. The NC molecules were diffused into the GAP-ETPE molecules, resulting in better compatibility between the GAP-ETPE and the NC. The high concentration of Bu-NENA did not increase the tendency of the phases to separate. Instead, it improved the molecular interactions between the GAP-ETPE and the NC, enhancing their compatibility.

The Han curve (lgG''–lgG' curve) [28,29] of the GAP-ETPE/NC/Bu-NENA blend adhesives was also studied to better understand the phase behavior. This curve is very sensitive to the morphological changes of multi-component or multiphase polymers and can be used to distinguish phase separations. Generally, the Han curves of homogeneous polymers at different temperatures can be superimposed to obtain a main curve, and there is no temperature dependence. For heterogeneous polymers (where phase separation exists in the polymer), the Han curves at different temperatures cannot overlap onto a main curve, which is temperature-dependent [28,29]. Studying the Han curve of a blended adhesive can help understand its phase change. Figure 9 shows the Han curve of the GAP-ETPE/NC/Bu-NENA-blended adhesive system at different temperatures. In Figure 9, the Han curves of the blending adhesive N30 are separated from each other at different temperatures and cannot be overlapped to form a single curve. There is an obvious temperature dependence, indicating a phase separation of GAP-ETPE and NC in N30. After adding Bu-NENA, the Han curves of the GAP-ETPE/NC/Bu-NENA system at different temperatures still do not overlap into a single curve, but the Han curves overlap at high temperatures. Above 120 °C, the Han curves of E-N-N-10 and E-N-N-15 overlap, and the Han curves of E-N-N-20 overlap at 110 °C when the Bu-NENA content is greater than 20%. The Han curves of Bu-NENA-blended adhesive systems almost overlapped at different temperatures. This is evidence that Bu-NENA compatibilized GAP-ETPE and NC in blending systems.

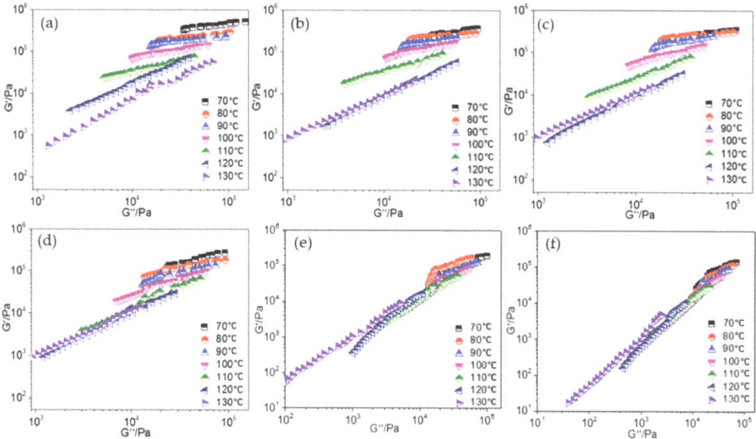

Figure 9. Han curve of the ETPE/NC/NENA-blended adhesive system: (**a**) N30, (**b**) E-N-N-10, (**c**) E-N-N-15, (**d**) E-N-N-20, (**e**) E-N-N-25, (**f**) E-N-N-30.

Combining the results of SEM imaging and Han curve analysis, Bu-NENA not only effectively plasticized NC and ETPE but also improved the compatibility between NC and GAP-ETPE. The phase separation in the adhesive was weakened, and the molecular mobility was enhanced. Finally, the toughness of the adhesive was significantly enhanced.

4. Conclusions

Based on the advantages of GAP-ETPE and the urgent need for high-safety propellants in current weapon systems, this paper proposed a novel high-strength solid propellant adhesive system using GAP-ETPE as the adhesive skeleton, Bu-NENA as the energetic plasticizer, and NC as the reinforcing agent. The crosslinking density, mechanical properties, and DSC showed that with increased Bu-NENA content, the crosslinking density and mechanical strength of GAP-ETPE/NC/Bu-NENA blends decreased monotonically, and the elongation at break first decreased and then increased. The T_g of the blending system decreased, and the T_g of NC was reduced to 120–140 °C. The results showed that this binder system is expected to give the propellant better mechanical and safety properties. These results provide a reference for the structure design, forming process, and parameter selection of a high-performance GAP-based green solid propellant.

Author Contributions: Y.L. contributed to the conception of the study; J.Z. and Z.W. performed the experiment; J.Z. performed the data analyses and wrote the manuscript; S.S. contributed significantly to analysis and manuscript preparation. All authors helped perform the analysis with constructive discussions and reviewed the manuscript. All authors have read and agreed to the published version of the manuscript.

Funding: This research received no external funding.

Institutional Review Board Statement: This study did not require ethical approval.

Data Availability Statement: Not applicable.

Conflicts of Interest: The authors declare no conflict of interest.

References

1. Tan, H.M. *The Chemistry and Technology of Solid Rocket Propellant*; Beijing Institute of Technology Press: Beijing, China, 2015; pp. 114–420.
2. Kubota, N. *Propellants and Explosives*; Wiley-VCH: Weinheim, Germany, 2006; pp. 69–112.
3. Kubota, N. Energetics of HMX-Based Composite Modified Double-Base Propellant Combustion. *J. Propuls. Power* **1999**, *15*, 759–762. [CrossRef]
4. An, C.W.; Li, F.S.; Wang, J.Y.; Guo, X.D. Surface Coating of Nitroamine Explosives and Its Effects on the Performance of Composite Modified Double-Base Propellants. *J. Propuls. Power* **2015**, *28*, 444–448. [CrossRef]
5. Nair, U.R.; Gore, G.M.; Sivabalan, R.; Divekar, C.N.; Asthana, S.N.; Singh, H. Studies on Advanced CL-20-based Composite Modified Double-Base Propellants. *J. Propuls. Power* **1971**, *20*, 952–955. [CrossRef]
6. Yano, Y.; Kubota, N. Combustion of HMX-CMDB Propellants (II). *Propellants Explos. Pyrotech.* **2010**, *11*, 1–5. [CrossRef]
7. Asthana, S.N.; Athawale, B.K.; Singh, H. Impact, Friction, Shock Sensitivities and DDT Behaviour of Advanced CMDB Propellants. *Def. Sci. J.* **1989**, *39*, 99–107. [CrossRef]
8. Choudhari, M.K.; Dhar, S.S.; Shrotri, P.G.; Singh, H. Effect of High Energy Materials on Sensitivity of Composite Modified Double Base (CMDB) Propellant System. *Def. Sci. J.* **2013**, *42*, 253–257. [CrossRef]
9. Gautam, G.K.; Pundlik, S.M.; Joshi, A.D.; Mulage, K.S.; Singh, S.N. Study of Energetic Nitramine Extruded Double-Base Propellants. *Def. Sci. J.* **2013**, *48*, 235–243. [CrossRef]
10. Reese, D.A.; Groven, L.J.; Son, S.F. Formulation and Characterization of a New Nitroglycerin-Free Double Base Propellant. *Propellants Explos. Pyrotech.* **2014**, *39*, 205–210. [CrossRef]
11. Matečić Mušanić, S.; Sućeska, M. Artificial ageing of double base rocket propellant. *J. Therm. Anal. Calorim.* **2009**, *96*, 523–529. [CrossRef]
12. Chin, A.; Ellison, D.; Poehlein, S. Investigation of the Decomposition Mechanism and Thermal Stability of Nitrocellulose/Nitroglycerine Based Propellants by Electron Spin Resonance. *Propellants Explos. Pyrotech.* **2007**, *32*, 117–126. [CrossRef]
13. Qi, X.F.; Yan, N.; Yan, Q.L.; Li, Y.F. Mesoscopic dynamic simulations on the phase structure of nitrocellulose/Plasticisers blends. *Chin. J. Explos. Propellants* **2017**, *40*, 101–107.
14. Lu, X.; Ding, L.; Chang, H.; Zhu, Y.L.; Wang, G.Y. Study on thermal hazard of plasticisers Bu-NENA and NG. *J. Solid Rocket Technol.* **2020**, *43*, 756–762.

15. Wang, Z.; Zhang, T.F.; Luo, Y.J. Effect of hard-segment content on rheological properties of glycidyl azide polyol-based energetic thermoplastic polyurethane elastomers. *Polym. Bull.* **2016**, *73*, 1–10. [CrossRef]

16. Badgujar, D.M.; Talawar, M.B.; Zarko, V.E.; Mahulikar, P.P. New directions in the area of modern energetic polymers: An overview. *Combust. Explos. Shock Waves* **2017**, *53*, 371–387. [CrossRef]

17. Gayathri, S.; Reshmi, S. Nitrato Functionalized Polymers for High Energy Propellants and Explosives: Recent Advances. *Polym. Adv. Technol.* **2017**, *28*, 1539–1550. [CrossRef]

18. Jarosz, T.; Stolarczyk, A.; Wawrzkiewicz-Jalowiecka, A.; Pawlus, K.; Miszczyszyn, K. Glycidyl azide polymer and its derivatives-versatile binders for explosives and pyrotechnics: Tutorial review of recent progress. *Molecules* **2019**, *24*, 4475. [CrossRef]

19. Wang, Z.; Zhang, T.F.; Zhao, B.B.; Luo, Y.J. Effect of different ratio nitrocellulose (NC) on morphology, rheological and mechanical properties of glycidyl azide polymer based energetic thermoplastic elastomer (ETPE)/NC blends. *Polym. Int.* **2016**, *66*, 705–711. [CrossRef]

20. *GB/T528-1998*; Rubber, Vulcanized or Thermoplastic—Determination of Tensile Stress-Strain Properties. Standardization Administration of the People's Republic of China: Beijing, China, 1998.

21. Wang, L.X.; Xue, J.Q.; He, W.G.; Zhou, J.Y.; Yu, H.C.; Shang, B.K. Properties and application of BuNENA energetic plasticizer. *Chem. Propellants Polym. Mater.* **2014**, *12*, 1–22.

22. Hu, Y.W.; Zheng, Q.L.; Zhou, W.L.; Xiao, L.Q. Preparation and properties of GAP-ETPE/NC blends. *Energ. Mater.* **2016**, *24*, 331–335.

23. Qi, X.F.; Zhang, X.H.; Guo, X.; Zhang, W.; Chen, Z.; Zhang, J.P. Experiment and simulation of NENA on NC plasticizing. *J. Solid Rocket Technol.* **2013**, *36*, 516–520.

24. Ding, H.Q.; Xiao, L.Q.; Jian, X.X.; Zhou, W.L. Study on GAP thermoplastic elastomer blending toughening Nitrocotton. *J. Solid Rocket Technol.* **2012**, *35*, 495–498.

25. Liu, Z.R.; Zhang, P. Influence of plasticizers on dynamic mechanical properties of double-base propellants. *J. Solid Rocket Technol.* **2005**, *28*, 276–279.

26. Cadogan, D.F.; Howick, C.J. Plasticizers. In *Kirk-Othmer Encyclopedia of Chemical Technology*; John Wiley & Sons, Inc.: Hoboken, NJ, USA, 2000.

27. Campbell, D.; Cumming, A.S.; Marshall, E.J. Development of Insensitive Rocket Propellants Based on Ammonium Nitrate and PolyNIMMO. In Proceedings of the Insensitive Munitions Symposium Technology, Williamsburg, VA, USA, 6–9 June; American Defense Preparedness Association: Arlington, VA, USA, 1994; pp. 229–239.

28. Han, C.D.; Chuang, H. Criteria for rheological compatibility of polymer blends. *J. Appl. Polym. Sci.* **2010**, *30*, 4431–4454. [CrossRef]

29. Han, C.D.; Yang, H. Rheological behavior of compatible polymer blends. I. Blends of poly(styrene-Co-acrylonitrile) and poly(ε-caprolactone). *J. Appl. Polym. Sci.* **1987**, *33*, 1199–1220. [CrossRef]

 *polymers*

Article

Preparation and Properties of Hydrophobic Polyurethane Based on Silane Modification

Yuxian Ma [1,2], Minghui Zhang [1,2], Wenhao Du [1,2], Shixiong Sun [1,2], Benbo Zhao [1,2,*] and Yuan Cheng [1,*]

1   School of Chemistry and Chemical Engineering, North University of China, Taiyuan 030051, China
2   Dezhou Industrial Technology Research Institute, North University of China, Dezhou 253034, China
*   Correspondence: 20170027@nuc.edu.cn (B.Z.); chengyuan@nuc.edu.cn (Y.C.); Tel.: +86-13703581570 (B.Z.); +86-13934606966 (Y.C.)

Abstract: Waterborne coatings have obtained more and more attention from researchers with increasing concerns in environmental protection, and have the advantages of being green, environmentally friendly and safe. However, the introduction of hydrophilic groups leads to lower hydrophobicity and it is difficult to meet the requirements of complex application environments. Herein, we proposed an optimization approach of waterborne polyurethane (WPU) with vinyl tris(β-methoxyethoxy) silane (A172), and it was found that the surface roughness, mechanical properties, thermal stability and water resistance of WPU will be increased to a certain extent with the addition of A172. Moreover, the hydrophobicity of the coating film is best when the silicon content is 10% of the acrylic monomer mass and the water contact angle reaches 100°, which could exceed two-thirds of the research results in the last decade. Therefore, our study can provide some theoretical basis for the research of hydrophobic polyurethane coatings.

Keywords: silane coupling agent; waterborne polyurethane; acrylate; hydrophobic modification

Citation: Ma, Y.; Zhang, M.; Du, W.; Sun, S.; Zhao, B.; Cheng, Y. Preparation and Properties of Hydrophobic Polyurethane Based on Silane Modification. *Polymers* **2023**, *15*, 1759. https://doi.org/10.3390/polym15071759

Academic Editor: Sándor Kéki

Received: 7 March 2023
Revised: 24 March 2023
Accepted: 28 March 2023
Published: 31 March 2023

1. Introduction

In recent years, considering environmental protection issues, waterborne coatings have been developed rapidly to reduce the emission of volatile organic compounds (VOCs). Among them, thanks to other resins and its ease of modification, waterborne polyurethane (WPU) could be conveniently modified by various methods to enhance its performance due to its excellent compatibility with other resins, which makes WPU widely used. However, the addition of hydrophilic chain extenders in the synthesis process resulted in the poor water resistance of WPU, which is the main obstacle in its application [1,2].

The most common modification methods to improve the properties of WPU include the introduction of acrylate monomers, silicones, organofluorine and nanomaterials [3–6]. Among them, the approach by introducing acrylates in WPU obtain the most attention, which could combine the advantages of acrylates and WPU to optimize the performance of coatings. However, due to the limitation of the flexible chain segment of acrylate, it still fails to meet the expected requirements (such as the water resistance always being unsatisfactory), so it is often combined with other modified methods to improve the comprehensive performance of the product [7]. Silicones are commonly used to improve this situation because of their outstanding properties such as high water resistance and heat resistance [8,9]. However, the strong hydrophobicity of silicone makes polyurethane difficult to emulsify in water and the phase separation degree will gradually increase due to the different solubility parameters of silicone and polyurethane, which may result in a decrease of emulsion stability. In order to overcome this contradiction, a variety of methods have been used to design the structure of silicone, such as the introducing of silane coupling agents, polyhedral low-polysilioxane (POSS), polar groups and polyether modification [10–13]. Among them, polyether modification and the introduction of polar group modification are both aimed at enhancing the water resistance of WPU by increasing the polarity of silicone, which often fail to achieve the expected effect, while modification with POSS always faces with issue of expensive raw materials and a complicated preparation process, the result of which is that it is difficult to use in most cases.

As a relatively common modification method, silane coupling agent contains both organic groups and inorganic silicon atoms, the latter of which have a smaller specific surface area and lower surface energy and can endow coatings with better hydrophobicity, high temperature resistance, etc. [14]. The main methods of modification with silane coupling agents include end modification, side-chain modification and synergistic modification [15]. For example, Gurunathan [16] and Gaddam et al. [17], respectively, modified WPU by end modification with 3-aminopropyltrimethoxysilane (APTMS) and 3-aminopropyltriethoxysilane (APTES), which contain only one amino group. During the reaction, the polyurethane prepolymer is directly sealed and the alkoxy group at the other end are hydrolyzed and condensed to form a cross-linked structure and improve the hydrophobicity of the material. However, considering that the use of end modification will limit the addition of silane, which will improve the performance of WPU to a limited extent [18], the side-chain modification method has been gradually considered to modify the WPU. Fu [19] and Lei et al. [20], respectively, modified WPU by side-chain modification with 3-mercaptopropyltrimethoxysilane and (3-(2-aminoethyl)aminopropyl)trimethoxysilane (AEAPTMS). Compared to the end modification, this method has a higher cross-link density and a larger relative molecular mass, which makes the WPU have better hydrophobicity and lower water absorption. In addition, in order to improve the comprehensive performance and versatility of the coatings, the synergistic modification has also been gradually developed [21]. For example, Zhao [22] and Yan et al. [23], respectively, modified WPU by synergistic modification with N-(2-aminoethyl)-3-aminopropyltriethoxysilane (AATS) and 3-glycidylethoxypropyltrimethoxysilane (GPTMS) together with APTMS. Zhang [24] used 3-(2-aminoethylamino)propyl dimethoxymethylsilane (KH-602) as a silane coupling agent to modify WPU with cyclophosphamide (PNMPD) containing double hydroxyl groups in synergistic side groups to obtain products with high fire resistance. From the above, we can see that the silane coupling agent not only has a wide range of applications, but also has a variety of pathways to improve the performance of WPU, which has a very far-reaching development prospect.

Among them, the stability and chemical reactivity of ethylene-based silane compounds make them suitable for use as silane coupling agents, and they are generally welcomed by producers and users because of the easy availability of synthetic raw materials, their simpler preparation methods and lower production costs. In the performance of modified polyurethane coatings, on the one hand, the addition of silane provides a hydrophobic surface with low surface energy. On the other hand, the hydrolysis of siloxane produces organosilanol, which will not exist stably, and it will further condense to generate silicon–oxygen–silicon bonds (Si-O-Si), leading to an increase in the chemical cross-linking point within the emulsion and an increase in cross-link density, which has an enhanced effect on the dense surface layer of the coating film and ultimately improves the thermal stability and mechanical properties of the coating film and the water resistance of the polymer. For example, Jing et al. [25] modified WPU with vinyltriethoxysilane (A151) by free radical polymerization reaction. It was found that the water contact angle of WPU was increased to 97° with the addition of A151, and the mechanical properties and solvent resistance of the coating film were also improved to some extent. Li et al. [26] prepared acrylate emulsions with self-crosslinking structures by introducing vinyltrimethoxysilane (A171) to address the problem of the poor water resistance of acrylate emulsions. It was found that the addition of A171 enhanced the water resistance and thermal stability of the resin to a certain extent. However, this modification also has certain defects, such as the hydrolytic condensation reaction of alkoxy groups being difficult to control [27], which makes it difficult for WPU to form and stabilize storage.

Compared to the alkoxy (methoxy, ethoxy, etc.) in general coupling agents, the 2-methoxyethoxy of vinyltris(β-methoxyethoxy)silane (A172) has a larger molecular weight.

The larger alkoxy structure makes its hydrolysis slower and more stable, which can solve the problem of the difficult-to-control hydrolysis condensation reaction of silane groups to some extent. Moreover, as a more special organosilane coupling agent, it has better solubility compared with other types of silane coupling agents because of its own molecular structure containing an ether-type structure, which makes its surface coating modification of hydrophilic inorganic fractions more adequate and efficient.

To the best knowledge of authors, no researchers have reported the use of A172 to modify WPU. Therefore, a series of silane-modified waterborne polyurethane acrylates (SWPUA) with hydrophobic groups in the chain segments, which were used in hydrophobic wood coatings fields, were prepared by selecting vinyltris(β-methoxyethoxy) silane (A172) as the modified monomer and combining it with acrylates for cross-linking modification of WPU, and the effects of different additions of A172 on the polyurethane properties were investigated. As expected, the polymerization emulsions could exhibit an excellent overall performance and the best hydrophobicity when the silicon content was 10% of the acrylic monomer mass, which could exceed two-thirds of the research results in the last decade. Accordingly, we believe that our study can provide some theoretical basis for the research of hydrophobicity of wood coatings.

2. Materials and Methods

2.1. Materials

Toluene diisocyanate (TDI, Tech) was purchased from Gansu Yinguang Chemical Industry Base Co. (Baiyin, China) Polyether diol (DL1000, Tech) was provided by Shandong Blue Star Dongda Co. (Zibo, China) 2,2-Dihydroxymethylpropionic acid (DMPA, Tech) was provided by Shenzhen Golden Tenglong Industrial Co. (Shenzhen, China). 1,4-Butanediol (BDO, Tech) was produced by Jining Huakai Resin Co. (Jining, China). Dibutyltin dilaurate (DBTDL, CP) was purchased from Aladdin Reagent Co. (Shanghai, China). Triethylamine (TEA, AR) and *N,N*-dimethylformamide (DMF, AR) were provided by Tianjin Zhiyuan Chemical Reagent Co. (Tianjin, China) Methyl methacrylate (MMA, AR), hydroxyethyl acrylate (HEA, AR), and vinyltris(β-methoxyethoxy)silane (A172, AR) were produced by Shanghai Maclean Biochemical Technology Co. (Shanghai, China) Butyl acrylate (BA, AR) was purchased from Fuchen Chemical Reagent Co. (Tianjin, China) Potassium persulfate (KPS, AR) was provided by Tianjin Hengxing Chemical Reagent Manufacturing Co. (Tianjin, China) Trimethylolpropane triacrylate (TMPTA, AR) was produced by RYOJI Ryosei.

2.2. Synthesis

2.2.1. Synthesis of Waterborne Polyurethane Emulsion (WPU)

Firstly, a certain amount of TDI (34.2 g), DL-1000 (70.1 g) and DBTDL were added into a four-necked flask equipped with a thermometer, mechanical stirrer and reflux condenser at 90 °C until the -NCO content reached the theoretical value. Then, DMPA (6.0 g), BDO (5.7 g), and HEA (4.1 g) were added sequentially, and the reaction time was determined by measuring the -NCO content during all periods. After cooling to 40 °C, a measured amount of TEA was added to neutralize the reaction for 50 min. Finally, the WPU emulsion was obtained after emulsifying for 1 h with deionized water under vigorous stirring. The reaction equation of WPU was shown in Figure 1.

2.2.2. Synthesis of Silane-Modified Waterborne Polyurethane Acrylate Emulsion (SWPUA)

Firstly, A certain amount of MMA, BA, HEA, TMPTA, silane coupling agent A172, deionized water and WPU emulsion were added into a four-necked flask equipped with a thermometer, mechanical stirrer and reflux condenser. After 30 min, KPS (dissolved with deionized water) was added and stirred for 10 min, and then pre-emulsion was obtained by filtration. Then, a certain amount of deionized water was weighed into a four-necked flask, and after the temperature of the system was increased to 80 °C, the pre-emulsion was transferred to a constant pressure-dispensing funnel and the flow rate was controlled so that the pre-emulsion was added dropwise within 3 h. Subsequently, the system temperature was raised to 85 °C and kept warm for 1 h, and then the post-initiator was added dropwise to the four-neck flask within 20 min. Finally, it was kept warm for 1 h, and the SWPUA emulsion was obtained by cooling and filtering. The reaction equation of SWPUA is shown in Figure 2 and the ratio of raw materials is listed in Table 1.

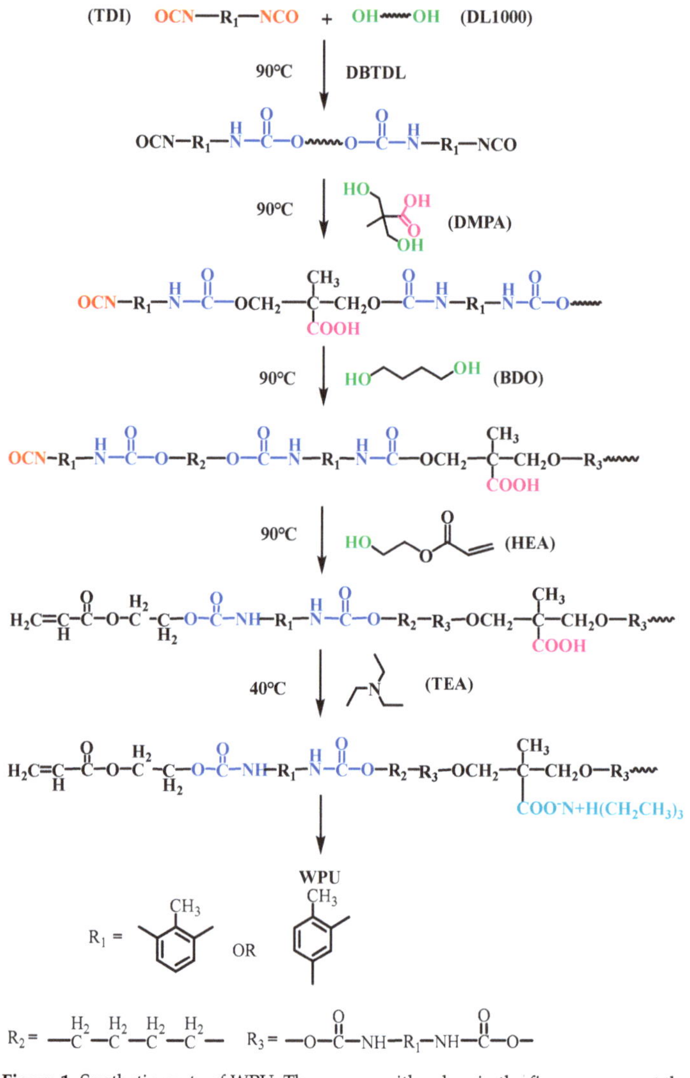

Figure 1. Synthetic route of WPU. The groups with colors in the figure represent the groups involved in the reaction. Red represents −NCO, green represents −OH, blue represents −NHCOO, purple represents −COOH, and cyan represents −COO⁻N⁺H(CH₂CH₃)₃.

2.2.3. Preparation of Coating Film

A certain amount of emulsion was weighed and spread into the PTFE plates, then the films were dried at room temperature for 24 h. At last, the required films were obtained after placed in a vacuum drying oven at 30 °C for 24 h.

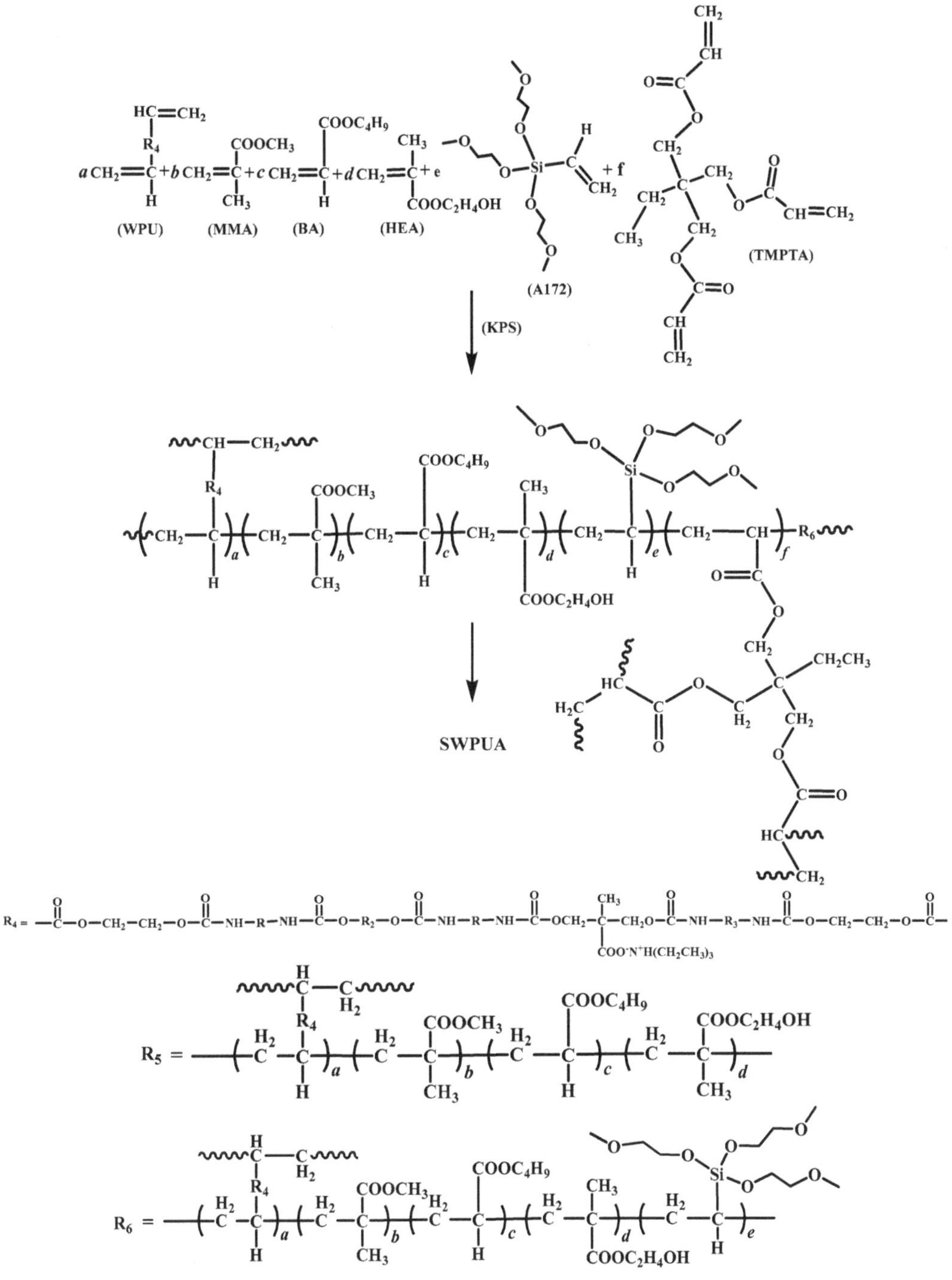

Figure 2. Synthetic route of SWPUA.

Table 1. The experimental formulation of SWPUA.

WPU	MMA	BA	HEA	TMPTA	A172	
g	g	g	g	g	ω_{A172}/%	g
100	39.9	18.6	1.5	1.2	0	0
100	39.9	18.6	1.5	1.2	5	3
100	39.9	18.6	1.5	1.2	10	6
100	39.9	18.6	1.5	1.2	15	9
100	39.9	18.6	1.5	1.2	20	12

2.2.4. Preparation of Paint Film

According to "GB/T1727.92 General Preparation Method of Paint Film" [28], it is prepared by brush coating method. First, the veneer was sanded smooth along the texture direction and the surface was cleaned of wood chips. Then, the emulsion was evenly applied to the veneer in the direction of the grain and dried naturally at room temperature for 24 h and then in a vacuum drying oven at 30 °C for 24 h. Ultimately, the paint film was obtained.

2.3. Characterization

Fourier transform infrared spectrometer (FTIR) was employed to characterize the chemical structure of WPU and SWPUA films at the attenuated total reflection (ATR) mode with a scan range of 4000~400 cm^{-1} and a resolution of 4 cm^{-1}.

The storage stability of WPU and SWPUA emulsions was tested in accordance with GB/T6753.3-1986 for a period of six months.

The centrifugal stability of WPU and SWPUA emulsions was tested by ultracentrifuge, and the centrifuge speed was set at 3000 r/min and the centrifugation time was 15 min.

A benchtop SEM (TM4000) was used to observe the longitudinal section of the coating film at a magnification of 500× and the surface morphology of the coating film at a magnification of 1000×.

The WPU and SWPUA emulsions were diluted to be 0.1 wt% solution with distilled water, and then the particle size and potential were measured with a nanoparticle size and zeta potential tester at 25 °C. Each sample was scanned three times, and the results were averaged.

A mechanical testing machine was used to perform the tensile test of the specimens, which were cut into a dumbbell shape and subjected to stress–strain measurements at a strain rate of 5 mm/min. At least three measurements were performed for each specimen.

The thermal performance analysis of the coating films was measured with a thermogravimetric analyzer. The test conditions were from 25 °C to 600 °C at a heating rate of 10 °C/min under N$_2$ atmosphere.

The contact angle of water on paint film was measured with a contact angle measuring instrument according to GB/T 30693-2014.

The prepared coating films were cut into pieces with size of 20 × 20 × 1 mm^3 and immersed into deionized water at 25 °C for 24 h. The water absorption of the film was calculated by the following formula:

$$\text{Water absorption} = \frac{m_1 - m_0}{m_0} \times 100\% \tag{1}$$

where m_0 is the mass of dried film and m_1 is the mass of the film after being put into the water for 24 h.

3. Results and Discussion

3.1. FT-IR Analysis

The FTIR spectra of WPU and SWPUA are shown in Figure 3. As can be seen from Figure 3, according to the FTIR spectra before and after the modification, the characteristic peaks of NH stretching vibration and C=O stretching vibration, respectively, appeared at 3440 cm^{-1} and 1730 cm^{-1}, and the absorption peak of 1240 cm^{-1} belonged to the asymmetric stretching vibration of C-O-C in the urethane group, and the stronger absorption peaks of -CH$_3$ and -CH$_2$- stretching vibration appeared at 2963 cm^{-1} and 2922 cm^{-1}, which together indicated that the urethane group was formed in the system. In addition, the characteristic absorption peak of -NCO disappeared at 2270 cm^{-1}, implying that the NCO groups have completely participated in the reaction.

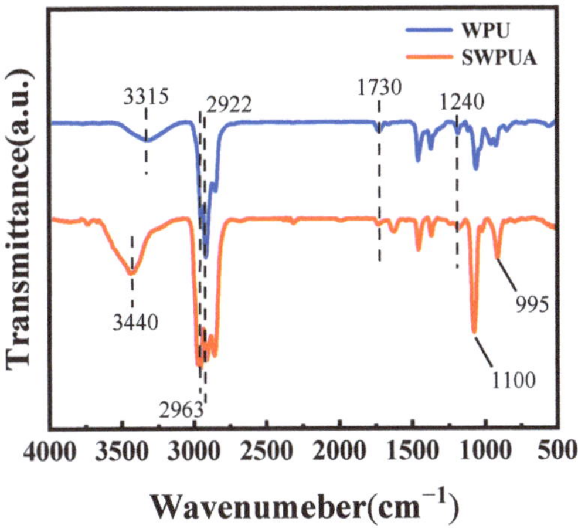

Figure 3. FT-IR spectra of WPU and SWPUA.

Compared with the spectra of WPU, it can be found that the absorption peak of Si-C stretching vibration appeared at 995 cm^{-1} in the spectra of SWPUA, and the absorption peak at 1240 cm^{-1} was enhanced and broadened, indicating that in addition to the C-O-C stretching vibration, a Si-O stretching vibration also appeared, and the characteristic absorption peak of C=C disappeared at 3000–3100 cm^{-1}, which indicated that free radical polymerization has occurred and the double bond had completely reacted, i.e., the silane coupling agent had been successfully introduced to the WPUA chain segment and the SWPUA emulsion had been prepared successfully [29]. In addition, it can be seen in the infrared spectrogram of SWPUA that stretching vibration absorption peak of Si-O-Si appeared at 1100 cm^{-1}, which indicates the hydrolysis of alkoxy group in silane coupling agent A172. The organosilanol was produced because of the hydrolysis of the silane coupling agent, which will not exist stably, and it will further condense to generate siloxane bond (Si-O-Si).

3.2. Surface Morphology Analysis

The longitudinal images of WPU and SWPUA coating films at a SEM magnification of 500 and the surface images at a magnification of 1000 are shown in Figure 4.

As shown in Figure 4a$_1$,a$_2$, striated structures with more directional distribution appeared on the longitudinal section of WPU, while more scale-like structures appeared on the longitudinal section of SWPUA. Moreover, the scale-like structure of SWPUA was bigger and more disordered. The reason for this structural change may be that the addition of silicon destroys the regularity of the polyurethane structure and increases the degree of microphase separation of polyurethane [30].

Comparing the surface images of WPU Figure 4b$_1$ and SWPUA Figure 4b$_2$, it can be found that the surface of SWPUA coating film was rougher than WPU, which showed finer graininess on the image. This is because of the migration of silicone with low surface tension to the surface of the coating film after the addition of A172 [31]. Besides, the results were in good agreement with the hydrophobic results of the coating film.

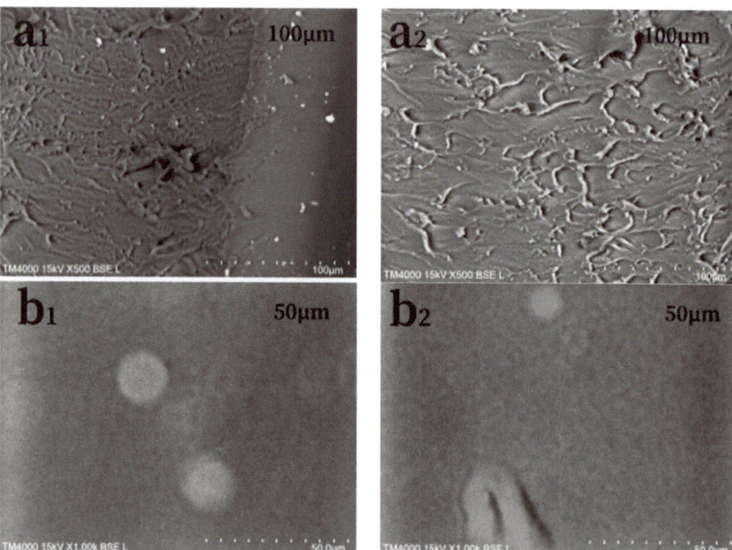

Figure 4. TM4000 images of WPU longitudinal section (**a₁**), WPU surface (**b₁**), SWPUA longitudinal section (**a₂**), SWPUA surface (**b₂**).

3.3. Particle Size and Stability Analysis

Compared to the unmodified emulsion, the particle size of the emulsion increased with the addition of A172, which could be found from Figure 5. This is because the addition of A172 increases the molecular weight of the system, and the increase of molecular weight leads to an increase of the particle size of the system [32]. In addition, with the increase of the content of A172, the average particle size of the emulsion showed a trend of first decreasing and then increasing. This is because the introduction of unsaturated bonded siloxanes enhances the interaction between the polyurethane and acrylate chains, resulting in tighter intermolecular connections. As a result, the average particle size of the emulsion particles was reduced. However, when the grafting amount of the silane coupling agent reached more than 10% of the acrylate monomer mass, the particle size and PDI tended to increase. The reason for this phenomenon may be that an increase in vinyl siloxane content leads to an increase in the unsaturation of the system, which leads to cross-linking and agglomeration between emulsion particles, resulting in an increase in particle size and irregular distribution, ultimately leading to an increase in the average particle size and PDI of the emulsion with an increase in siloxane content [33].

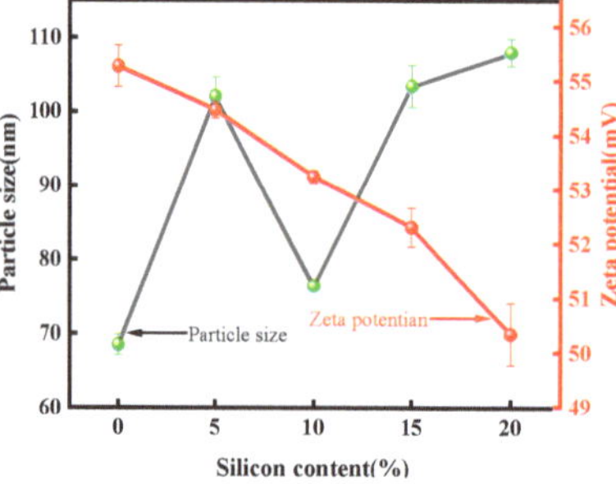

Figure 5. Particle size and zeta potential of SWPUA emulsion with different silicon content.

It can be found from Figure 5 that the zeta potential showed a gradual decrease trend with the addition of certain silane coupling agents. This is because during the emulsion polymerization process, the hydrophobic silane coupling agent enters into the emulsion particles along with the acrylate monomer to form the inner core, which leads to the swelling of the emulsion particles and the reduction of the density of ionic groups on the surface of the polymer particles, eventually making the repulsive force between the polymer particles decrease. Moreover, the zeta potential is a measure of the strength of the interaction between the particles, so the zeta potential gradually decreases.

As shown in Table S1 (in Supplementary Materials), when the addition of A172 was higher than 10% of the initial acrylate monomer mass, the storage stability of the emulsion decreased to less than 6 months, and precipitation occurred by centrifugation. On the one hand, this is because with the increase of the content of silane coupling agent, the cross-linking point and the cross-linking degree in the polymer molecular system increase, which leads to the gradual change of molecular chain from linear to bulk, and the intermolecular movement changes from intermolecular forces to chemical bonds, so the resistance of movement becomes larger. Therefore, the stability decreased and precipitation occurred. On the other hand, during the long-term storage process, the emulsion particles settle under the action of gravity and form a concentrated layer at the bottom of the container, which makes a reduction in the spacing of the emulsion particles, and some of the particles cross the potential barrier and become unstable and coalesce. In addition, the larger the particle size of the emulsion, the faster the settling velocity of the emulsion particles, which is detrimental to the storage stability of the emulsion.

3.4. Mechanical Performance Analysis

The stress–strain curves of polyurethane coating films with different silane coupling agent additions are shown in Figure 6a, and their maximum tensile strength and elongation at break with silicon content are shown in Figure 6b. The specific data is shown in Table 2.

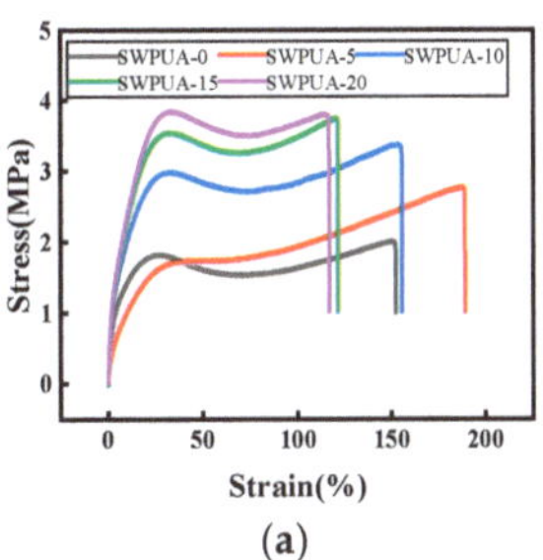

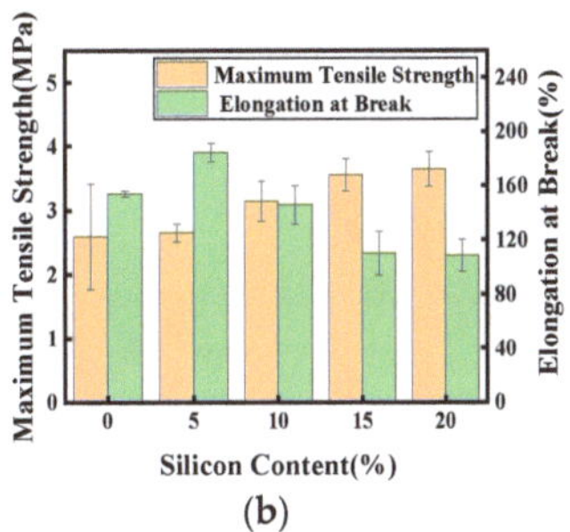

Figure 6. (a) Stress–Strain curves of SWPUA samples with different silicon content; (b) Maximum tensile strength and Elongation at break of SWPUA samples with different silicon content.

Table 2. The maximum tensile strength and elongation at break of SWPUA coatings with different silicon content.

Sample	SWPUA-0	SWPUA-5	SWPUA-10	SWPUA-15	SWPUA-20
Silicon content/%	/	5	10	15	20
Maximum tensile strength/MPa	2.59 ± 0.82	2.65 ± 0.14	3.14 ± 0.31	3.56 ± 0.25	3.64 ± 0.27
Elongation at break/%	154 ± 2	184 ± 7	146 ± 14	110 ± 16	108 ± 12

According to Figure 6 and Table 2, The mechanical properties of WPU were improved to some extent with the addition of A172, and the tensile strength of the films gradually increased with the increase of the content of silane coupling agent A172, and the elongation at break showed a trend of increasing and then decreasing [25–28]. The reason why the tensile strength of SWPUA increases is probably because when A172 comes into WPU molecule, it can effectively increase the cross-link density of the system, which makes the intermolecular force enhanced and restricts the movement of the molecular chain. When small contents of A172 were added, the elongation at break increased. This is probably because the addition of a small amount of A172 increases the degree of microphase phase separation between hard and soft segments, which makes the intermolecular arrangement less tight. Therefore, the elongation at break increases. However, when the contents of A172 continues to increase, the elongation at the break decreased because when the silane coupling agent was added to a certain concentration, the cross-linking effect plays a major role compared to the microphase separation effect, which would restrict the movement of molecular chains; thus, the elongation at the break tended to decrease [34–37].

3.5. TG Analysis

The TG curves and DTG curves of waterborne polyurethane with different silicon content are shown in Figure 7a,b, respectively. As seen in Figure 7, the thermal decomposition process of waterborne polyurethane film was mainly divided into three stages. In general, the first degradation stage in the temperature ranged from 0 to 200 °C, which was mainly caused by the decomposition of small molecules and the evaporation of water. The second stage of decomposition ranged from 250 to 350 °C, which was mainly the decomposition of the carbamate and urea bonds in the hard chain segment. Finally, the last stage ranged from 350 to 450 °C, which was the decomposition of ether bond and silica bond in the soft chain sections.

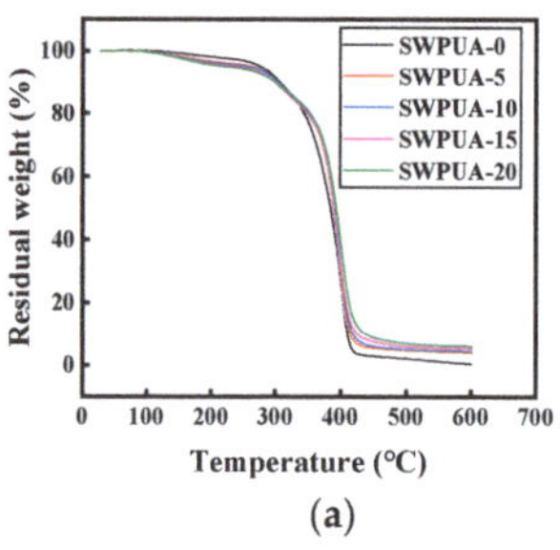
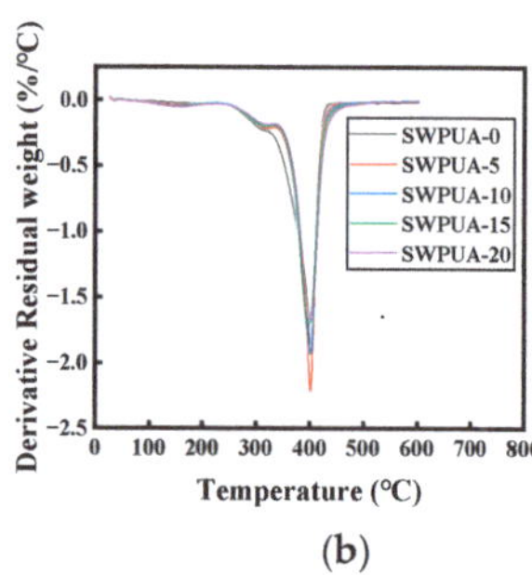

Figure 7. (**a**) Weight loss curves of SWPUA samples with different silicon content; (**b**) Derivative of weight loss curves of SWPUA samples with different silicon content.

As shown in the figure, the thermal stability of WPU was improved to some extent with the addition of A172. With the introduction of silane coupling agent, the temperature at 50% mass loss (T50%) and final residue mass of the modified specimens all increased. Moreover, the final residue mass of coating films increased with the increase of silicon content. On the one hand, because of the introduction of silicon–oxygen bond, the bond energy of Si-O bond is higher than C-C and C-O bonds [38,39], which makes it more difficult to break (Si-O bond energy is 550 KJ/mol, C-C or C-O bond energy is 340 KJ/mol). On the other hand, because of the double bond of A172, the increase in the content of the double bond leads to the increase in the degree of cross-linking, and the chemical bonds between polyurethane and polyacrylate molecules also enhance the intermolecular interactions. The synergistic effect of the two ultimately leads to the increasing of thermal stability of the coating film. Therefore, the addition of silicon can effectively increase the thermal stability of the WPU coating film.

3.6. Water Absorption and Water Contact Angle Analyses

As seen in Figure 8a, with the addition of silane coupling agent, the water resistance of WPU coating film has been improved to a certain extent, which is intuitively shown by the decrease of the water absorption rate. With the increase of the silicon content, the water absorption of the coating film showed a trend of gradual decrease. The reason for this is that the addition of silane coupling agent leads the increase of the cross-link density of the system, which makes it difficult for hydrophilic groups to move to the surface of the film and further prevents water molecules from penetrating into the interior of the film; thus, the water absorption of the coating film decreases.

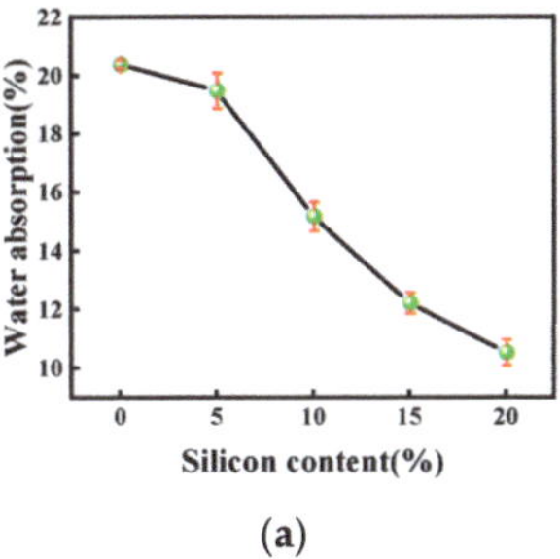
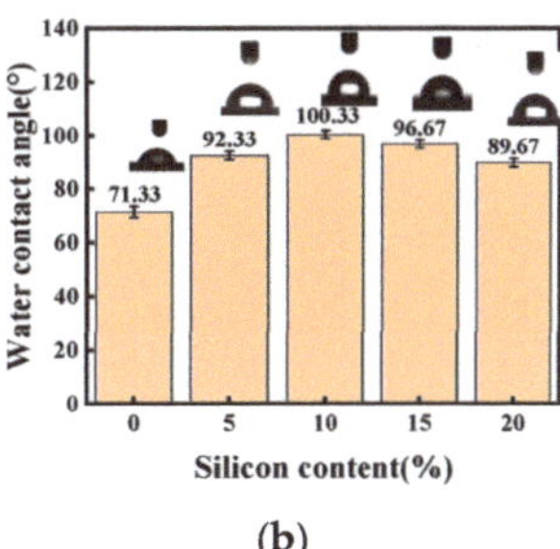

(a)   (b)

Figure 8. (**a**) Water absorption of SWPUA samples with different silicon content; (**b**) Water contact angle of SWPUA samples with different silicon content.

As shown in Figure 8b, with the addition of A172, the hydrophobicity of WPU was improved to a certain extent, which was reflected by the increase of water contact angle. With an increase in the content of A172 in SWPUA film from 0 to 20%, the water contact angle of SWPUA coating film first increased obviously and then decreased. Among them, when the A172 content was added at 10% of the acrylic monomer mass, the water contact angle reached a maximum value of 100°. The reason why the water contact angle tends to increase first is because of the migration of silicon to the surface of the coating film and the outward orientation of the side-chain silanes of the polymer, which reduces the surface energy of the coating film and corresponds to the formation of a hydrophobic surface with low surface energy; thus, the water contact angle and hydrophobicity increase. However, when the silicon content exceeds 10% of the acrylic monomer mass, the water contact angle decreases, because as the silicon content continues to increase, on the one hand, the effective enrichment of silicon on the surface reaches the maximum value, a critical micelle concentration appears and the reduction of surface energy gradually tends to level off, so the water contact angle no longer increases. On the other hand, when the degree of cross-linking of the polymer increases to a certain extent, the potential resistance effect is obvious, which will further reduce the polymeric grafting of polyurethane and acrylate resin; thus, the water contact angle decreases [40,41]. Therefore, the addition of a certain amount of A172 can enhance the surface hydrophobicity of coatings to a certain extent. In the last decade, there are few studies on enhancing the hydrophobicity of polyurethane with silicon modification, and some data on hydrophobicity studies of polyurethane in the last decade at home and abroad were briefly listed in Figure 9 [12,18,20,25,30,31,42–55].

It can be found from Figure 9 that the optimal addition amounts of silicon for each system varied from one research work to another because of the different original formulation and the type of silicon modification. It can be found that the water contact angle of the modified coating film is mostly concentrated between 80 and 100° so far. It is generally accepted that when the water contact angle is greater than 90°, the coating film is considered to have good hydrophobicity. Therefore, in comparison with other research works, it is clear that our findings are at an advanced level, exceeding two-thirds of the research results.

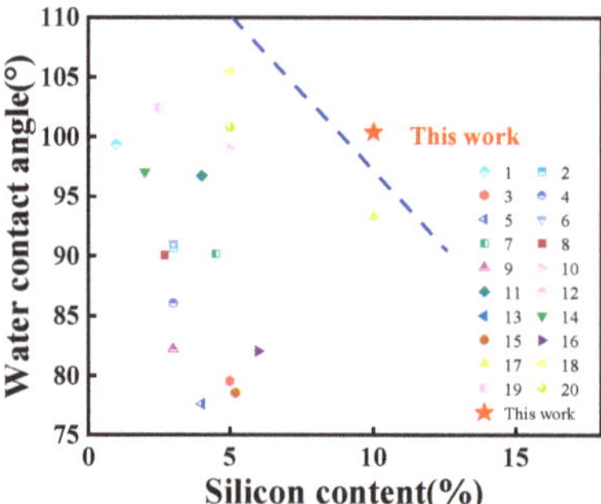

Figure 9. The maximum water contact angle of different research systems.

4. Conclusions

In this study, a series of modified waterborne polyurethane acrylates (SWPUA) with different silicon contents were prepared by cross-linking the WPU with vinyl tris(β-methoxyethoxy)silane (A172), which will be applied to the use of hydrophobic wood coatings. The results showed that the addition of A172 led to different degrees of improvement of hydrophobicity, water resistance, mechanical properties and thermal stability of WPU, among which the improvement of hydrophobicity was the most significant. With the addition of silane coupling agent, the particle size of emulsion showed a trend of increasing, then decreasing and then increasing again, and the Zeta potential showed a decreasing trend. It was found that the emulsion modified with A172 could be stably stored for 6 months. Scanning electron microscopy showed that the addition of silane coupling agent increased the surface roughness and the degree of microphase separation of the coating film. After the coating film formation, the tensile test and TG test data showed that the mechanical properties and thermal stability of the coating were improved. The water absorption test revealed that the water absorption of the coating film gradually decreased. The hydrophobicity test of SWPUA showed that the water contact angle of the film tended to increase first and then slightly decrease. Under the condition of stable storage of the emulsion, when the addition of A172 was 10% of the acrylate mass, the water contact angle of SWPUA film was increased from 71° to 100°, which could exceed two-thirds of the research results at home and abroad in the last decade. The results showed that without affecting the storage stability of the emulsion, the addition of A172 could solve the problem of poor hydrophobicity of waterborne polyurethane to some extent. In addition, it can improve the comprehensive performance of polyurethane to a certain extent.

Supplementary Materials: The following supporting information can be downloaded at: https://www.mdpi.com/article/10.3390/polym15071759/s1, Table S1. Particle size, polydispersity index (PDI), Zeta potential and stability of SWPUA emulsion with different silicon content. Table S2. The serial numbers in the radar diagram and their corresponding references.

Author Contributions: Conceptualization, Y.M., M.Z. and W.D.; methodology, Y.M. and M.Z.; validation, Y.M. and M.Z.; formal analysis, W.D., S.S. and B.Z.; investigation, Y.M., M.Z., W.D., S.S., B.Z. and Y.C.; resources, B.Z. and S.S.; data curation, Y.M. and M.Z.; writing—original draft preparation, Y.M., W.D. and S.S.; writing—review and editing, W.D., S.S., B.Z. and Y.C.; visualization, W.D., S.S. and B.Z.; supervision, W.D., S.S., B.Z. and Y.C.; project administration, S.S. and B.Z.; funding acquisition, B.Z. and Y.C. All authors have read and agreed to the published version of the manuscript.

Funding: This research was funded by the Industry–University–Research Innovation Fund of the Science and Technology Development Center of the Ministry of Education (Grant No. 2021DZ018) and Scientific and Technological Innovation Programs of Higher Education Institutions in Shanxi (Grant No. 2020L0322).

Institutional Review Board Statement: Not applicable.

Informed Consent Statement: Not applicable.

Data Availability Statement: The data presented in this study are available on request from the corresponding author.

Conflicts of Interest: The authors declare no conflict of interest.

References

1. Zhan, X.; Chen, J.; Yang, Z.; Wu, G.; Kong, Z. Superhydrophobic film from silicone-modified nanocellulose and waterborne polyurethane through simple sanding process. *Int. J. Biol. Macromol.* **2023**, *232*, 123431. [CrossRef]
2. Li, F.; Liang, Z.; Li, Y.; Wu, Z.; Yi, Z. Synthesis of waterborne polyurethane by inserting polydimethylsiloxane and constructing dual crosslinking for obtaining the superior performance of waterborne coatings. *Compos. Part B Eng.* **2022**, *238*, 109889. [CrossRef]
3. Tsupphayakorn-aek, P.; Suwan, A.; Tulyapitak, T.; Saetung, N.; Saetung, A. A novel UV-curable waterborne polyurethane-acrylate coating based on green polyol from hydroxyl telechelic natural rubber. *Prog. Org. Coat.* **2022**, *163*, 106585. [CrossRef]
4. Bakhshandeh, E.; Sobhani, S.; Croutxé-Barghorn, C.; Allonas, X.; Bastani, S. Siloxane-modified waterborne UV-curable polyurethane acrylate coatings: Chemorheology and viscoelastic analyses. *Prog. Org. Coat.* **2021**, *158*, 106323. [CrossRef]
5. Wang, X.; Cui, Y.; Wang, Y.; Ban, T.; Zhang, Y.; Zhang, J.; Zhu, X. Preparation and characteristics of crosslinked fluorinated acrylate modified waterborne polyurethane for metal protection coating. *Prog. Org. Coat.* **2021**, *158*, 106371. [CrossRef]
6. Hong, C.; Zhou, X.; Ye, Y.; Li, W. Synthesis and characterization of UV-curable waterborne Polyurethane–acrylate modified with hydroxyl-terminated polydimethylsiloxane: UV-cured film with excellent water resistance. *Prog. Org. Coat.* **2021**, *156*, 106251. [CrossRef]
7. Hwang, H.-D.; Kim, H.-J. Enhanced thermal and surface properties of waterborne UV-curable polycarbonate-based polyurethane (meth) acrylate dispersion by incorporation of polydimethylsiloxane. *React. Funct. Polym.* **2011**, *71*, 655–665. [CrossRef]
8. Wang, L.; Shen, Y.; Lai, X.; Li, Z.; Liu, M. Synthesis and properties of crosslinked waterborne polyurethane. *J. Polym. Res.* **2011**, *18*, 469–476. [CrossRef]
9. Liu, J.; Jiao, X.; Cheng, F.; Fan, Y.; Wu, Y.; Yang, X. Fabrication and performance of UV cured transparent silicone modified polyurethane–acrylate coatings with high hardness, good thermal stability and adhesion. *Prog. Org. Coat.* **2020**, *144*, 105673. [CrossRef]
10. Ren, L.; Yu, S.; Niu, Q.; Qiang, T. Construction of amphiphilic comb-like waterborne polyurethane to balance emulsion stability and film hydrophobicity. *Prog. Org. Coat.* **2022**, *173*, 107197. [CrossRef]
11. Song, H.; Zhang, Q.; Zhang, Y.; Wang, Y.; Zhou, Z.; Zhang, P.; Yuan, B. Waterborne polyurethane/3-amino-polyhedral oligomeric silsesquioxane (NH 2-POSS) nanocomposites with enhanced properties. *Adv. Compos. Hybrid Mater.* **2021**, *4*, 629–638. [CrossRef]
12. Zheng, G.; Lu, M.; Rui, X. The effect of polyether functional polydimethylsiloxane on surface and thermal properties of waterborne polyurethane. *Appl. Surf. Sci.* **2017**, *399*, 272–281. [CrossRef]
13. Wang, Y.; An, Q.; Yang, B. Synthesis of UV-curable polyurethane acrylate modified with polyhedral oligomeric silsesquioxane and fluorine for iron cultural relic protection coating. *Prog. Org. Coat.* **2019**, *136*, 105235. [CrossRef]
14. Karna, N.; Joshi, G.M.; Mhaske, S. Structure-property relationship of silane-modified polyurethane: A review. *Prog. Org. Coat.* **2023**, *176*, 107377. [CrossRef]
15. Deng, F.; Qin, S.; Liu, N.; Xu, W. The Synergistic Effect of Terminal and Pendant Fluoroalkyl Segments on Properties of Polyurethane Latex and Its Film. *Coatings* **2022**, *12*, 1271. [CrossRef]
16. Gurunathan, T.; Chung, J.S. Synthesis of aminosilane crosslinked cationomeric waterborne polyurethane nanocomposites and its physicochemical properties. *Colloids Surf. A Physicochem. Eng. Asp.* **2017**, *522*, 124–132. [CrossRef]
17. Gaddam, S.K.; Kutcherlapati, S.R.; Palanisamy, A. Self-cross-linkable anionic waterborne polyurethane–silanol dispersions from cottonseed-oil-based phosphorylated polyol as ionic soft segment. *ACS Sustain. Chem. Eng.* **2017**, *5*, 6447–6455. [CrossRef]
18. Lyu, J.; Xu, K.; Zhang, N.; Lu, C.; Zhang, Q.; Yu, L.; Feng, F.; Li, X. In situ incorporation of diamino silane group into waterborne polyurethane for enhancing surface hydrophobicity of coating. *Molecules* **2019**, *24*, 1667. [CrossRef]
19. Fu, C.; Yang, Z.; Zheng, Z.; Shen, L. Properties of alkoxysilane castor oil synthesized via thiol-ene and its polyurethane/siloxane hybrid coating films. *Prog. Org. Coat.* **2014**, *77*, 1241–1248. [CrossRef]
20. Lei, L.; Zhang, Y.; Ou, C.; Xia, Z.; Zhong, L. Synthesis and characterization of waterborne polyurethanes with alkoxy silane groups in the side chains for potential application in waterborne ink. *Prog. Org. Coat.* **2016**, *92*, 85–94. [CrossRef]
21. Zhang, Y.; Lin, R.; Shi, Y.; Li, H.; Liu, Y.; Zhou, C. Synthesis and surface migration of polydimethylsiloxane and perfluorinated polyether in modified waterborne polyurethane. *Polym. Bull.* **2019**, *76*, 5517–5535. [CrossRef]
22. Zhao, H.; Huang, D.; Hao, T.H.; Hu, G.H.; Ye, G.B.; Jiang, T.; Zhang, Q.C. Synthesis and investigation of well-defined silane terminated and segmented waterborne hybrid polyurethanes. *New J. Chem.* **2017**, *41*, 9268–9275. [CrossRef]
23. Wu, Y.; Du, Z.; Wang, H.; Cheng, X. Synthesis of aqueous highly branched silica sol as underlying crosslinker for corrosion protection. *Prog. Org. Coat.* **2017**, *111*, 381–388. [CrossRef]
24. Zhang, P.; Fan, H.; Tian, S.; Chen, Y.; Yan, J. Synergistic effect of phosphorus–nitrogen and silicon-containing chain extenders on the mechanical properties, flame retardancy and thermal degradation behavior of waterborne polyurethane. *RSC Adv.* **2016**, *6*, 72409–72422. [CrossRef]

25. Jing, R.; Wen, Y.; Yu, L.; Zhu, P. Synthesis and Properties of Vinyl Silane Coupling Agent Modified Waterborne Polyurethane. *Leather Sci. Eng.* **2015**, *25*, 45–49.
26. Liu, Y.; Wei, M. Synthesis and Characterization of Pure Acrylic Emulsion Modified with Silane Coupling Agent. *Paint. Coat. Ind.* **2019**, *42*, 6–9.
27. Jiao, C.; Shao, Q.; Wu, M.; Zheng, B.; Guo, Z.; Yi, J.; Zhang, J.; Lin, J.; Wu, S.; Dong, M.; et al. 2-(3, 4-Epoxy) ethyltriethoxysilane-modified waterborne acrylic resin: Preparation and property analysis. *Polymer* **2020**, *190*, 122196. [CrossRef]
28. GB/T1727.92; General Preparation Method of Paint Film. Standardization Administration of the People's Republic of China: Beijing, China, 1992.
29. Zhang, F.; Liu, W.; Liang, L.; Liu, C.; Wang, S.; Shi, H.; Xie, Y.; Yang, M.; Pi, K. Applications of hydrophobic α, ω-bis (amino)-terminated polydimethylsiloxane-graphene oxide in enhancement of anti-corrosion ability of waterborne polyurethane. *Colloids Surf. A Physicochem. Eng. Asp.* **2020**, *600*, 124981. [CrossRef]
30. Rao, Z.; Yan, H.; Tao, W.; Liu, C.; Jian, G.; Zhou, Y.; Chen, H.; Yang, M. Effects of aminopropyl-terminated polydimethylsiloxane on structure and properties of waterborne polyurethane-acrylate. *Prog. Org. Coat.* **2023**, *174*, 107314. [CrossRef]
31. Zhang, Y.; Lin, R.; Li, H.; Shi, Y.; Liu, Y.; Zhou, C. Investigation of the migration ability of silicone and properties of waterborne polyurethanes. *Int. J. Adhes. Adhes.* **2020**, *102*, 102654. [CrossRef]
32. Li, Q.; Guo, L.; Qiu, T.; Xiao, W.; Du, D.; Li, X. Synthesis of waterborne polyurethane containing alkoxysilane side groups and the properties of the hybrid coating films. *Appl. Surf. Sci.* **2016**, *377*, 66–74. [CrossRef]
33. Zhang, D.; Liu, J.; Li, Z.; Shen, Y.; Wang, P.; Wang, D.; Wang, X.; Hu, X. Preparation and properties of UV-curable waterborne silicon-containing polyurethane acrylate emulsion. *Prog. Org. Coat.* **2021**, *160*, 106503. [CrossRef]
34. Yu, X.; Xiong, Y.; Zhou, C.; Li, G.; Ren, S.; Li, Z.; Tang, H. Polyurethane with tris (trimethylsiloxy) silyl propyl as the side chains: Synthesis and properties. *Prog. Org. Coat.* **2020**, *142*, 105605. [CrossRef]
35. Chen, R.S.; Chang, C.J.; Chang, Y.H. Study on siloxane-modified polyurethane dispersions from various polydimethylsiloxanes. *J. Polym. Sci. Part A Polym. Chem.* **2005**, *43*, 3482–3490. [CrossRef]
36. Li, C.Y.; Chen, J.H.; Chien, P.C.; Chiu, W.Y.; Chen, R.S.; Don, T.M. Preparation of poly (IPDI-PTMO-siloxanes) and influence of siloxane structure on reactivity and mechanical properties. *Polym. Eng. Sci.* **2007**, *47*, 625–632. [CrossRef]
37. Zong, J.; Zhang, Q.; Sun, H.; Yu, Y.; Wang, S.; Liu, Y. Characterization of polydimethylsiloxane–polyurethanes synthesized by graft or block copolymerizations. *Polym. Bull.* **2010**, *65*, 477–493. [CrossRef]
38. Xu, C.A.; Qu, Z.; Tan, Z.; Nan, B.; Meng, H.; Wu, K.; Shi, J.; Lu, M.; Liang, L. High-temperature resistance and hydrophobic polysiloxane-based polyurethane films with cross-linked structure prepared by the sol-gel process. *Polym. Test.* **2020**, *86*, 106485. [CrossRef]
39. Kim, M.-G.; Jo, K.-I.; Kim, E.; Park, J.-H.; Ko, J.-W.; Lee, J.H. Preparation of polydimethylsiloxane-modified waterborne polyurethane coatings for marine applications. *Polymers* **2021**, *13*, 4283. [CrossRef]
40. Stefanović, I.S.; Džunuzović, J.V.; Džunuzović, E.S.; Brzić, S.J.; Jasiukaitytė-Grojzdek, E.; Basagni, A.; Marega, C. Tailoring the properties of waterborne polyurethanes by incorporating different content of poly (dimethylsiloxane). *Prog. Org. Coat.* **2021**, *161*, 106474. [CrossRef]
41. Sui, Z.; Li, Y.; Guo, Z.; Zhang, Q.; Xu, Y.; Zhao, X. Preparation and properties of polysiloxane modified fluorine-containing waterborne polyurethane emulsion. *Prog. Org. Coat.* **2022**, *166*, 106783. [CrossRef]
42. He, Z.; Xue, J.; Ke, Y.; Luo, Y.; Lu, Q.; Xu, Y.; Zhang, C. Enhanced Water Resistance Performance of Castor Oil—Based Waterborne Polyurethane Modified by Methoxysilane Coupling Agents via Thiol-Ene Photo Click Reaction. *J. Renew. Mater.* **2022**, *10*, 591. [CrossRef]
43. Zhang, S.; Chen, Z.; Guo, M.; Bai, H.; Liu, X. Synthesis and characterization of waterborne UV-curable polyurethane modified with side-chain triethoxysilane and colloidal silica. *Colloids Surf. A Physicochem. Eng. Asp.* **2015**, *468*, 1–9. [CrossRef]
44. Sun, D.; Miao, X.; Zhang, K.; Kim, H.; Yuan, Y. Triazole-forming waterborne polyurethane composites fabricated with silane coupling agent functionalized nano-silica. *J. Colloid Interface Sci.* **2011**, *361*, 483–490. [CrossRef] [PubMed]
45. Fu, C.; Hu, X.; Yang, Z.; Shen, L.; Zheng, Z. Preparation and properties of waterborne bio-based polyurethane/siloxane cross-linked films by an in situ sol–gel process. *Prog. Org. Coat.* **2015**, *84*, 18–27. [CrossRef]
46. Li, Y.; Zhao, T.; Qu, X.; Ding, H.; Li, F. Synthesis of waterborne polyurethane modified by nano-SiO_2 silicone and properties of the WPU coated RDX. *China Pet. Process. Petrochem. Technol.* **2015**, *17*, 39–45.
47. Du, Y.; Yang, Z.; Zhou, C. Study on waterborne polyurethanes based on poly (dimethyl siloxane) and perfluorinated polyether. *Macromol. Res.* **2015**, *23*, 867–875. [CrossRef]
48. Jiang, W.; Dai, A.; Zhou, T.; Xie, H. Hybrid polysiloxane/polyacrylate/nano-SiO_2 emulsion for waterborne polyurethane coatings. *Polym. Test.* **2019**, *80*, 106110. [CrossRef]
49. Christopher, G.; Kulandainathan, M.A.; Harichandran, G. Biopolymers nanocomposite for material protection: Enhancement of corrosion protection using waterborne polyurethane nanocomposite coatings. *Prog. Org. Coat.* **2016**, *99*, 91–102. [CrossRef]
50. Du, Y.; Wang, Z.; Zhou, T.; Cai, X.; Ren, X. Synthesis of silane coupling agent modified high solid waterborne polyurethane. *China Synth. Resin Plast.* **2012**, *29*, 16–20.
51. Wang, C.; Wang, Q.; Dai, Z.; Xu, G. Research of Waterborne Polyurethane Modified With Silane Coupling Agent KH-602. *Appl. Technol.* **2010**, *31*, 23–27.

52. Allauddin, S.; Narayan, R.; Raju, K. Synthesis and properties of alkoxysilane castor oil and their polyurethane/urea–silica hybrid coating films. *ACS Sustain. Chem. Eng.* **2013**, *1*, 910–918. [CrossRef]
53. Mahmoudi, M.; Javaherian Naghash, H. Synthesis and characterization of a novel hydroxyl-terminated polydimethylsiloxane for application in the silicone-modified acrylic-grafted polyester resins. *J. Adhes. Sci. Technol.* **2015**, *29*, 1341–1359. [CrossRef]
54. Yu, Q.; Pan, P.; Du, Z.; Du, X.; Wang, H.; Cheng, X. The study of cationic waterborne polyurethanes modified by two different forms of polydimethylsiloxane. *RSC Adv.* **2019**, *9*, 7795–7802. [CrossRef] [PubMed]
55. Ding, X.; Wang, X.; Zhang, H.; Liu, T.; Hong, C.; Ren, Q.; Zhou, C. Preparation of waterborne polyurethane-silica nanocomposites by a click chemistry method. *Mater. Today Commun.* **2020**, *23*, 100911. [CrossRef]

Article

Study on GAP Adhesive-Based Polymer Films, Energetic Polymer Composites and Application

Siyuan Wu, Xiaomeng Li *, Zhen Ge and Yunjun Luo

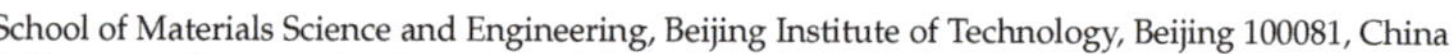

School of Materials Science and Engineering, Beijing Institute of Technology, Beijing 100081, China
* Correspondence: xm.lee@bit.edu.cn; Tel.: +86-13651278705

Abstract: To lay the foundation for environmentally friendly energetic polymer composites, GAP (glycidyl azide polymer) adhesive-based polymer films with different curing parameter R (mol ratio of hydroxyl/isocyanate) and energetic polymer composites with different RDX contents were studied. GAP/TDI (toluene diisocyanate)/GLY(glycerol) was selected as the adhesive system. The tensile strength and elongation at the break of the polymer film with R = 2.2, was 14.34 MPa and 176.86%, respectively, as observed by an AGS-J electronic universal testing machine. A relatively complete cross-linking network and high hydrogen bonding interaction were observed by LF-NMR (low-field nuclear magnetic resonance, where the cross-linking density was 11.06×10^{-4} mol/cm^3) and FT-IR (fourier transform infrared spectroscopy, where the carbonyl bonding ratio was 64.84%). Forty percent RDX(hexogen) was added into the adhesive system. The tensile strength was 4.65 MPa, and the elongation at the break was 78.49%; meanwhile, the heat of the explosive was 2.87 MJ/kg, and the residue carbon rate was only 2.47%. The tensile cross-sections of energetic polymer composites were observed by SEM (Scanning electron microscopy).

Keywords: adhesive; polymer film; energetic polymer composites; mechanical properties

Citation: Wu, S.; Li, X.; Ge, Z.; Luo, Y. Study on GAP Adhesive-Based Polymer Films, Energetic Polymer Composites and Application. *Polymers* **2023**, *15*, 1538. https://doi.org/10.3390/polym15061538

Academic Editor: Anton M. Manakhov

Received: 15 February 2023
Revised: 14 March 2023
Accepted: 17 March 2023
Published: 20 March 2023

1. Introduction

Energetic polymer composites, including solid propellant, polymer bonded explosive, gun propellants, and combustible cartridges, have been widely used as the power and damaged sources of missiles, torpedoes, guns, and explosives in the military weapon system due to their ultrahigh energy density and strong capacity for doing damage. However, when it comes to most of the energetic polymer composites, their polymer adhesives are not only non-energetic, but they also have low mechanical properties, leading to low energy density as well as mechanical properties [1].

GAP (glycidyl azide polymer), as an energetic adhesive, has the advantages of good burnout, high energy, high burning rate, and low sensitivity. N$_2$ and CO$_2$ are released after combustion, both of which are clean gases [2–5]. Therefore, it is very promising to prepare environmentally friendly energetic polymer composites [6]. Combustible cartridges, for example, are a kind of cartridge that can burn and provide energy. They have the advantages of being clean, light weight, providing energy supplies, not needing recycling like metal cartridges, reducing environmental pollution, and saving manpower and material costs [7–9]. It is a sustainable and environmentally friendly polymer. Therefore, we studied the use of the GAP adhesive system to prepare combustible components, such as combustible cartridges, and the design of the pouring process. However, GAP has poor mechanical properties due to the high steric hindrance of the azide group and the low number of atoms carried by the main chain, which is the common difficulty of GAP-based products, limiting the widespread use of GAP [10,11]. Researchers have explored several methods of overcoming this problem, such as adjusting the curing parameters, optimizing the cross-linking system, finding a suitable composite adhesive system, and adding reinforcing materials, etc. [9,12–16]. In this work, TDI was used as a curing agent, and glycerol was used as an adhesive. Additionally, we used glycerol as the three-functional cross-linking agent in the adhesive system, with a significant increase in the amount of glycerol, an increase in the cross-linking density of the system and a decrease in the molecular weight between the cross-linking points. The mechanical properties of GAP polymer film were optimized by changing the amount of glycerol and curing parameters. Subsequently, RDX was added to study the effects of the relative ratio of the adhesive system of RDX on the mechanical properties, the heat of the explosive, and the residue carbon rate. In this paper, the adhesive system was used to prepare combustible components such as combustible cartridges and cartridge boxes by the pouring process.

2. Materials and Methods

2.1. Materials

GAP (industrial grade, Mn = 3700) and dioctyl sebacate (DOS, industrial grade) were from the Luoyang Liming Chemical Research Institute (Luoyang, China). The GAP was dewatered in a vacuum drying oven at 80 °C for 24 h before use; glycerol (GLY, AR) was from Saran Chemical Technology Co., Ltd., Shanghai, China); toluene diisocyanate (TDI, industrial grade) was from Tianjin Guanghua Fine Chemical Research Institute (Tianjin, China); antifoaming agent (B-160, industrial grade) was from Guangdong Zhonglianbang Fine Chemical Co., Ltd., Guangdong, China; Dibutyltin dilaurate (DBTDL, AR) and triphenyl bismuth (TPB, AR) were from Beijing Chemical Plant (Beijing, China); Hexogen particles (RDX, 108 μm) were purchased from the North Huian Chemical Co., Ltd., Xi'an, China.

2.2. Preparation of Polymer Films

GAP and GLY were added to the beaker and stirred well. TDI, B-160, DBTDL and TPB (DBTDL: TPB = 1:3) were added to the above solution and mixed by a homogenizer keeping vacuum at 2500 r/m for 180 s to remove bubbles. The solution obtained was cast into a petri dish to cure for 7 days at 60 °C, as shown in Figure 1. In addition, the feeding ratio of all samples is shown in Table 1.

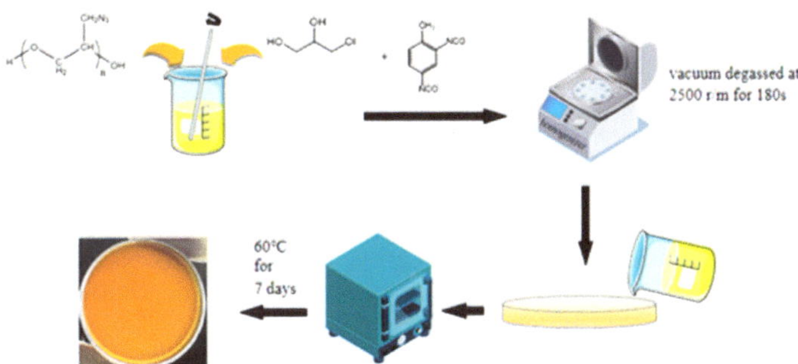

Figure 1. Schematic illustration for the preparation process of polymer films.

Table 1. Feeding ratio of polymer films with different R.

R	GAP/%	TDI/%	GLY/%
1.2	89.71	8.74	1.03
1.4	87.03	11.16	1.31
1.6	83.81	14.04	1.65
1.8	79.88	17.56	2.06
2.0	74.95	21.97	2.58
2.2	68.60	27.65	3.25
2.4	60.12	35.24	4.14

2.3. Preparation of Energetic Polymer Composites

RDX particles were added into the polymer matrix to prepare energetic polymer composite films, after determining the curing parameter (R, mol ratio of hydroxyl/isocyanate). Other steps were the same as above. The ambient humidity was 30% during the sample preparation process, because the ambient humidity can affect the sample curing if it is too high.

When the R = 1.0, the polymer film was not fully cured at 60 °C for 10 days; when the R = 2.6, although the sample could be cured at 60 °C for 1 day, the surface of the polymer film was uneven after curing because the sample was more sensitive to water. As a result, two samples with R = 1.0 and 2.6 could not be tested. Therefore, they will not be further discussed in the remainder of this text.

2.4. Characterization

The hardness of samples was tested by an LXD-A of the digital Shore A hardness tester (Shanghai Siwei Instrument Manufacturing Co., Ltd., Shanghai, China) at 25 °C. The chemical structure of samples was tested by Fourier transform infrared spectroscopy (FT-IR) by Nicolet 8700 infrared spectrometer (Thermo Fisher Scientific, Waltham, MA, USA). Test conditions: the test temperature was 25 °C, the number of scans was 32 times, the resolution was 4 cm^{-1}, and the scanning range was 500–4000 cm^{-1}. Static mechanical performance of samples was tested by an AGS-J electronic universal testing machine (Shimadzu Corporation, Japan), according to the method of GB/T528-1998. The test temperature was 25 °C, and the tensile rate was 100 mm/min. Low-field nuclear magnetic resonance (LF-NMR) of samples was tested by a VTMR20-010V-T nuclear magnetic resonance analyzer (Suzhou Niumai Technology Co., Ltd., Suzhou, China). Test conditions: The test temperature was 25 °C, and the number of accumulation times was 3. The heat of explosiveness of the samples was tested by an American Parr Company Parr6200 oxygen bomb calorimeter for testing, according to the GJB770B2005 gunpowder test method 701.1. The residue carbon rate was calculated by the proportion of the residual mass of the heat of the explosive sample. Scanning electron microscopy (SEM) was tested by A Hitachi S4800, Tokyo, Japan. Test conditions: acceleration voltage was 15 kv.

3. Results and Discussion

3.1. FT-IR of Polymer Films with Different R

Figure 2 shows the FT-IR spectrum of polymer films with different R; 2271 cm^{-1} and 2927 cm^{-1} are the characteristic absorption peaks of -NCO and C-H, and there is no obvious change with the increase in R, which indicates that both polyols and isocyanates have participated in the reaction; 2098 cm^{-1} is the -N$_3$ characteristic absorption peak of GAP; 3346 cm^{-1} is the N-H stretching vibration peak; 2919 cm^{-1} is the C-H asymmetric stretching vibration peak; 2864 cm^{-1} is the C-H symmetric stretching vibration peak; 1716 cm^{-1}, 1538 cm^{-1}, and 1279 cm^{-1} are the characteristic absorption peaks of amide I, II and III bands on carbamate; 1446 cm^{-1} is the -CH$_2$ bending vibration; 1353 cm^{-1} is the superposition of -CH bending vibration absorption peak; and -CN stretching vibration, the stretching vibration peak of C-O-C ether bond, is at 1109 cm^{-1}, which indicates that carbamate group was successfully synthesized and GAP polymer films were successfully prepared. The full FT-IR spectrum of the samples proves that the sample has been successfully synthesized.

Figure 3a shows the FT-IR spectrum of the carbonyl group of the polymer film with different R. To accurately analyze the hydrogen bond interaction of the polymer films, the second-order derivative spectrum was used to quantitatively identify the secondary structure of the carbonyl group; 1730 cm^{-1} and 1690 cm^{-1} were internal for the characteristic absorption peaks of free carbonyl and bonded carbonyl (Figure 3b). Gaussian function was used to fit the peak of the spectrum. After obtaining each sub-area, the ratio of its bonded carbonyl was calculated [17], and the calculated results are shown in Table 2.

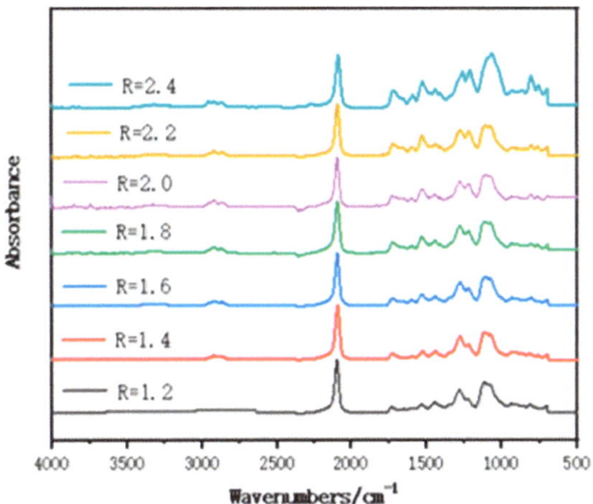

Figure 2. FT-IR spectrum of polymer films with different R.

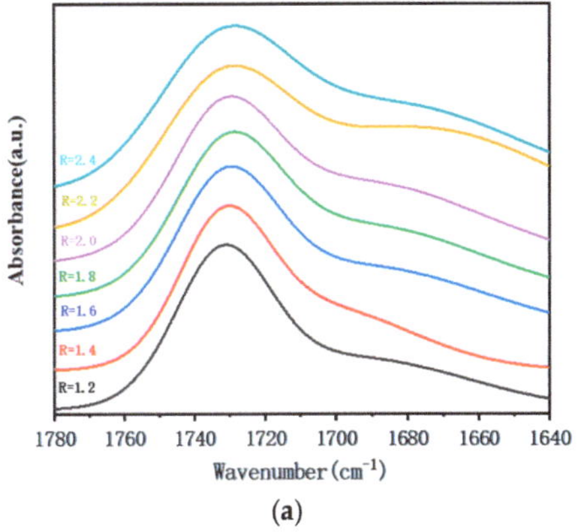

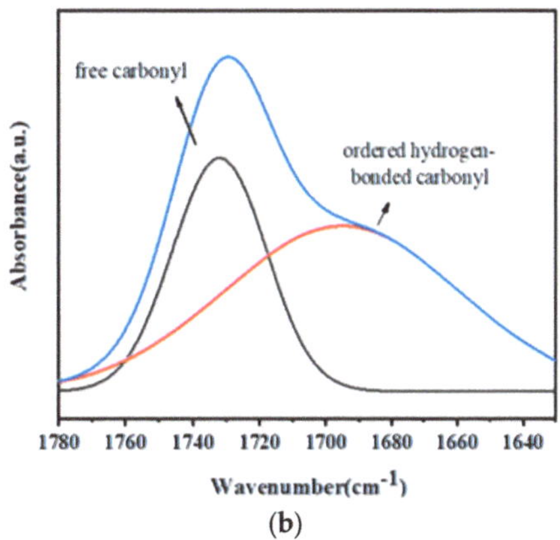

(a)   (b)

Figure 3. (a) FT-IR spectrum of carbonyl group of polymer films with different R; (b) FT-IR peak fitting spectrum of carbonyl group of polymer films.

Table 2. The bonded carbonyl ratio of GAP/TDI/GLY polymer films with different R.

R	Free-Carbonyl	Bonded-Carbonyl	Total-Carbonyl	Bonded-Carbonyl-Ratio (%)
1.2	2.0771	2.0006	4.0778	49.06
1.4	3.0331	3.1961	6.2291	51.31
1.6	9.9886	3.9343	6.9228	56.83
1.8	1.5847	2.1949	3.7796	58.07
2.0	0.6030	1.0923	1.6953	64.43
2.2	1.2901	2.3794	3.6695	64.84
2.4	8.8463	9.7253	16.572	58.69

Due to the formation of hydrogen bonds, the carbonyl peaks are split, and the FT-IR absorption peak positions of carbonyl groups that are involved in the formation of hydrogen bonds shift to lower wavenumbers. Thermodynamically, the structural units of the hard segment and the soft segment of the polymer films are incompatible or incompletely compatible, and the hard segments would aggregate with each other to produce dispersed micro domains and main micro phase separation. Micro phase separation not only played a role similar to filler reinforcement (inhibiting crack propagation), but also acted as a physical crosslink point for molecular chains, which is beneficial to improve the flammable cartridge mechanical properties. The hydrogen bonding between the hard segments in the polymer film is stronger, leading to a higher degree of micro phase separation.

The peak areas of free and bonded carbonyl groups are obtained from the sub-peaks, and the proportion of bonded carbonyl groups is shown in Table 2, which indicates that the proportion of bonded carbonyl groups of GAP/TDI/GLY polymer films increased first and then decreased, with the increase of R and the degree of micro phase separation also showing the same trend. The reason for that is: When the R is increased, the hard segments aggregate to form hard segment micro domains, and the degree of micro phase separation of the polymer film is enhanced due to the enhancement of hydrogen bonding between the hard segments. When R = 2.4, the mobility of the hard segment is restricted due to the high crosslink density in the polymer films, and the difficulty of formation of hydrogen bonds increases, resulting in a decrease in the degree of hard segment aggregation.

3.2. Crosslink Density of Polymer Films with Different R

The crosslinking density of polymer films was tested by low-field nuclear magnetic resonance (LF-NMR) to analyze the reason for enhancing mechanical tensile. As shown in Table 3, it was found that the crosslink density (ν_e) of the polymer film increased with the increase of R. This is mainly attributed to the following two factors: on the one hand, -NCO can react with the urethane group and form a chemical crosslinking point; on the other hand, more hard segments increase the degree of micro phase separation; these hard segments are also viewed as physical crosslinking points. The increase in crosslinking density can restrict the mobility of molecule chains, leading to an increase in mechanical tensile.

Table 3. Crosslink density and correction factor of polymer films with different R.

R	$\nu_e \times 10^{-4}$ (mol/cm^{-3})	ρ (g/cm^{-3})	E (MPa)	Mc (kg/cm^{-3})	A
1.2	1.115	1.299	0.055	1.165	−0.235
1.4	1.231	1.297	0.126	1.054	−0.237
1.6	1.302	1.295	0.226	0.995	−0.240
1.8	4.823	1.291	2.196	0.268	−0.361
2.0	8.033	1.288	6.709	0.160	0.409
2.2	11.06	1.283	17.001	0.116	3.157
2.4	12.84	1.277	71.228	0.099	20.815

The mechanical property of the polymer films is also related to the crosslink network structure. The actual crosslink network of the polymer film is not ideal, and there are other structural features [18,19] that are divided into the following cases: (1) the two ends of the same molecular chain are bonded to form a sealed circle; (2) only one end of the molecular chain enters the network structure, forming terminal defects; (3) temporary physical entanglement of molecular chains; (4) some groups between molecular chains form physical crosslinks due to hydrogen bonding. For these four cases, since quantitative calculation and statistics are not available, a correction factor (A) is introduced to represent the total contribution of these four effects to the mechanical properties of the crosslink network. Among the four cases, the former two cases cause defects in the crosslink network structure, resulting in a decrease in A, while the latter two cases will compensate for the elastic modulus of the crosslink network when subjected to external force, resulting in a larger A [17]. If A < 0, meaning that there are many network defects in the system; If A = 0, the compensation of elastic modulus by physical crosslink can offset the defects in the system; A > 0 means that the physical crosslink effect in the system is strong, and the crosslink network structure is more complete.

To study the crosslink network structure of the polymer film with different R, the relative molecular mass between the crosslink points (M$_c$) was calculated by the Formula (1).

According to Formula (2), we know that A is related to the shear modulus of samples. To calculate the shear modulus of samples, we first tested the elastic modulus of all samples and then calculated the shear modulus based on a relationship between elastic modulus and shear modulus (Formula (3)), when the strain of crosslink elastomer is very small. Based on experimental results and formulas, the correction factor (A) of all samples is shown in Table 3. As shown in Table 3, when R = 1.2–1.8, the A of the polymer film is negative, indicating that the system had many network structural defects, and the physical crosslink could not offset the negative effects of the network defects; when R = 2.0–2.4, the content of isocyanate groups increased in the system, the proportion of hydrogen bonds between molecular chains increased, and the probability of forming a cyclic structure also increased. These two interactions result in a gradual increase in the A value of the polymer films, implying that the physical crosslink effect is strong, and the structural integrity of the crosslink network is good. Therefore, with the increase of R, the cross-linking network structure of the polymer films was more complete, leading to an increase in tensile strength. In summary, the mechanical property of the polymer film is increased, which is due to high crosslink density and a more integrated molecular network structure.

$$v_e = \frac{\left(\ln(1 - v_2) + v_2 + \chi v_2^2\right)}{v\left(v_2^{1/3} - 2v_2/f\right)} = \frac{\rho_b}{M_c} \tag{1}$$

$$G = \rho RT / M_c + A \tag{2}$$

$$E = 3\rho RT / M_c + 3A \tag{3}$$

where v_e is the crosslink density of the elastomer; v_2 is the volume fraction of the rubber phase in the swollen elastomer; v is the molar volume of the solvent; χ is the interaction parameter between the adhesive and the solvent, and f is the adhesive network The functionality of; ρ_b is the density of the elastomer, M_c is the average molecular mass between the crosslink points; G is the shear modulus, E is the elastic modulus, M_c is the average molecular mass between the crosslink points, A is the correction factor, R is the thermodynamic constant, T is the temperature, ρ is the density.

3.3. Mechanical Property of Polymer Films with Different R

As a characterization of resistance to deformation, Shore A hardness can also be used to describe the curing degree of the curing system and measure whether the curing is complete. As shown in Figure 4a, other GAP/TDI/GLY polymer films can form a regular shape after curing for 48 h, and the hardness of the polymer films remained unchanged after curing for six days, indicating that the polymer film is completely cured. In the case of large-scale production, the samples could be demoulded after curing for 48 h and waiting for complete curing.

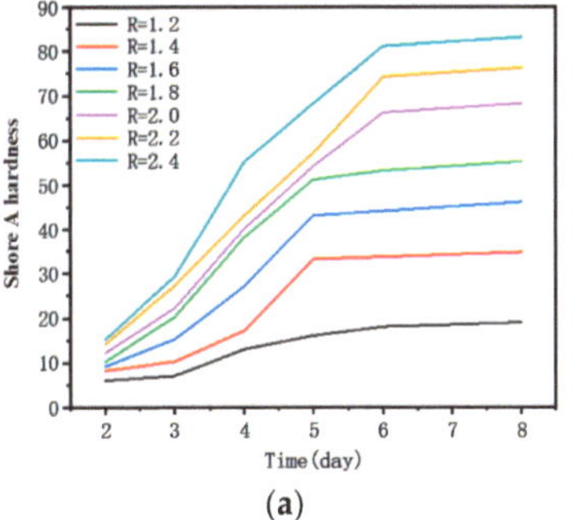

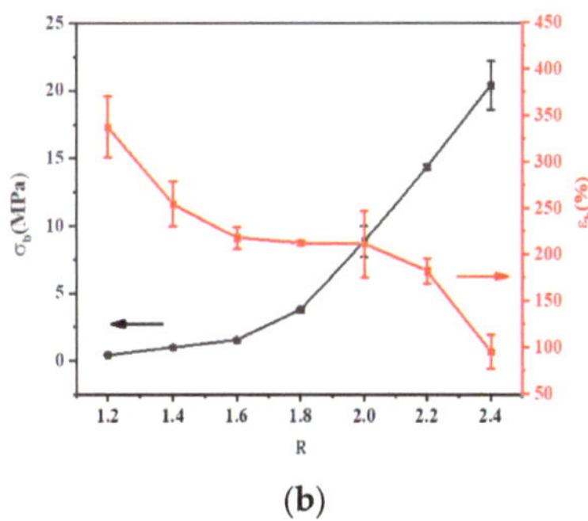

Figure 4. (**a**) Shore A hardness of the polymer films with different R; (**b**) curves of tensile strength and elongation at break of polymer films with different R.

As shown in Figure 4b, the tensile strength of polymer films increased with the increase of R, which is consistent with the results obtained from the proportion of infrared bonding. Due to the increase of R, the content of isocyanate and glycerol increased, and the content of hard segment and the cross-linking point in the polymer film increased, and the strength of the polymer film increased. However, as a three-functional cross-linking agent, the increase of the amount of glycerol significantly increased the cross-linking density of the system and reduced the molecular weight between the cross-linking points, resulting in the molecular chain being difficult to fully extend freely due to the increase of cross-linking density during the stretching process, while elongation at the break decreased. When R = 1.2–2.4, the tensile strength of polymer films increased from 0.43 MPa to 19.58 MPa, and the elongation at the break decreased from 337.51% to 94.98%, respectively. The higher tensile strength was better, yet a lower strength could be accepted if the elongation at break exceeds 30%. This is because the requirement of the propellant under normal circumstances is that the elongation at the break is higher than 30%. Since there is no specific requirement for the combustible element, this paper refers to the requirements of the propellant, considering that the tensile strength of polymer films would decrease after adding RDX. Therefore, herein we choose the range of R = 2.0, 2.2 and 2.4, the tensile strength of the polymer film is increased from 8.83 MPa to 19.58 MPa, and the elongation at the break decreased from 211.29% to 94.98%, respectively. Based on the above results, we find that when R = 2.0, 2.2, 2.4, the mechanical properties of samples are optimum. However, the polymer film with a higher R is more sensitive to water. Therefore, when R = 2.2, the polymer film is chosen and used as a polymer adhesive in energetic polymer composites in the next text. Some scholars have studied the method of improving GAP, and the optimized strength is generally achieved. For example, the strength of the film prepared by Xu and Ma et al. reached 1.5 MPa and 1.6 MPa, and the elongation reached 81.6%. Compared with the existing test results, the strength and elongation of the GAP film reported in this paper are much higher than the existing results [11,14,16].

3.4. The SEM of Tensile Cross-Section of Energetic Polymer Composites

The SEM images of tensile cross-section of the polymer films with different R were shown in Figure 5a–g. With R increases, the 'circles' appear in the section, due to the degree of phase separation in the polymer film increasing and the stripes after tensile fracture becoming more obvious. The overall sections were relatively smooth, indicating that the films obviously had brittle fractures under tensile action. The number of folds in the section increased with the increase of R, and the actual surface area of the fracture surface increased, so that a large amount of energy was dissipated through the generation of new surface.

In order to understand the dispersion of RDX in the adhesive system, the fracture surface of energetic polymer composite with 40% RDX was observed by SEM (Figure 5h). The tensile fracture surface was rougher, and the particle distribution was more uniform. However, the RDX particles were separated from the adhesive system, and some holes appeared. These holes were formed by the RDX particles distributed in the films falling off the fracture surface under the action of external force. Increasingly obvious folds appeared around the holes and particles, indicating that the RDX particles combined with the matrix can be used as stress concentration sites to transfer stress and dissipate energy, thereby reducing the crack growth rate.

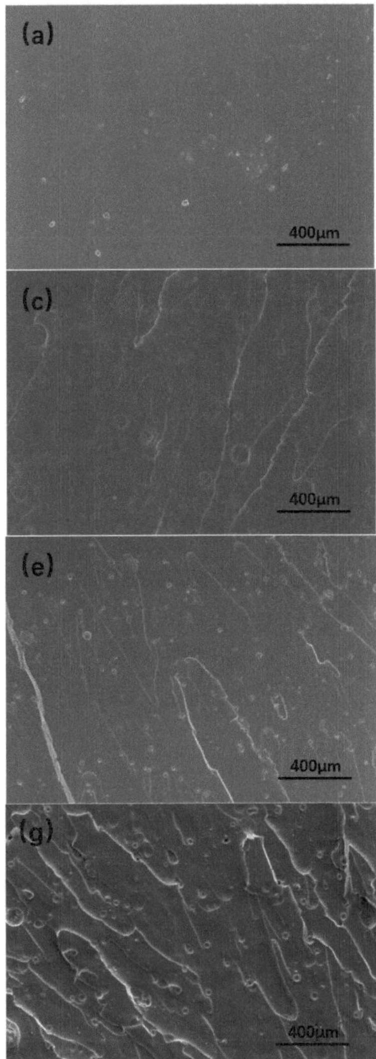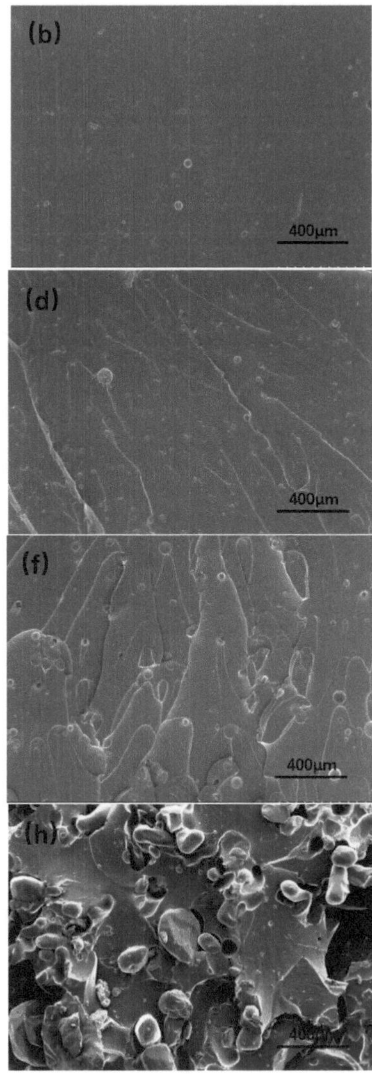

Figure 5. The SEM of tensile cross-section of polymer films with different R = 1.2 (**a**), 1.4 (**b**), 1.6 (**c**), 1.8 (**d**), 2.0 (**e**), 2.2 (**f**), 2.4 (**g**), and energetic polymer composites with 40% RDX with R = 2.2 (**h**).

3.5. Mechanical Property of Energetic Polymer Composites with Different RDX Contents

The mechanical properties of energetic polymer composites with different RDX particle contents were tested. The mechanical tensile and elongation at the breaks of polymer composites are shown in Figure 6. When the content of RDX particles is increased from 10 wt.% to 60 wt.%, the mechanical strength of polymer composites decreased from 8.44 MPa to 2.41 MPa, and the elongation at break decreased from 134.18% to 14.60%. This reason is attributed to two factors: one is that with the increases of RDX particle content, the content of polymer adhesive decreases; the other is that there is a poor interfacial binding force between polymer adhesive and RDX particles; these particles cannot constrain the movement of polymer chains, leading to a decrease in mechanical properties.

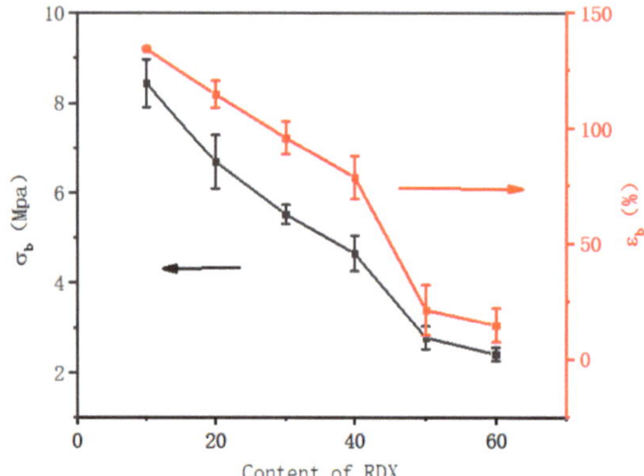

Figure 6. The curves of tensile strength and elongation at break of energetic polymer composites with different RDX contents.

3.6. The Heat of Explosive and Residual Carbon Rate of Energetic Polymer Composites with Different RDX Contents

To study key properties of energetic polymer composites, the heat of explosive and residual carbon rate with different RDX contents was tested—the results are shown in Table 4. It was observed that with the increase of RDX contents, the heat of the explosive of polymer composites increased, and residual carbon rate decreased, indicating that polymer composite with 60 wt.% RDX perform well.

Table 4. The heat of explosive and residue carbon rate of energetic polymer composites with different RDX contents.

RDX Content (%)	The Heat of Explosive (MJ/kg)	Residue Carbon Rate (%)
10	2.25	6.75
20	2.41	5.28
30	2.51	4.09
40	2.87	2.47
50	3.09	2.20
60	3.29	1.78

4. Conclusions

Energetic polymer composites were prepared with different R consisting of GAP used as the adhesive, TDI, and glycerol and its composites. With the increase of R, mechanical strength gradually increased from 0.43 MPa to 19.58 MPa, and elongation at break decreased from 337.51% to 94.98%. When the content of RDX particles increases with R = 2.2, the mechanical tensile and elongation at break of polymer composites both decrease from 8.44 MPa to 2.41 MPa, and from 134.18% to 14.60%, respectively. In addition, when the loading of the RDX content is 60 wt.% and R = 2.2, the values of the heat of explosive and residue carbon rate of polymer composites are 3.29 MJ/kg and 1.78%, respectively. GAP as a clean adhesive has been widely studied in aerospace, polymers, and other fields, as it is environmentally friendly. This work provides support for the research on improving the mechanical properties of eco-friendly energetic polymer composites.

Author Contributions: Conceptualization, S.W., X.L. and Z.G.; methodology, S.W., X.L. and Z.G.; validation, S.W., X.L. and Z.G.; formal analysis, S.W.; investigation, S.W.; resources, S.W.; data curation, S.W.; writing—original draft preparation, S.W.; writing—review and editing, S.W., X.L., Z.G. and Y.L.; visualization, S.W.; supervision, X.L.; project administration, S.W., X.L., Z.G. and Y.L.; funding acquisition, X.L. and Z.G. All authors have read and agreed to the published version of the manuscript.

Funding: This research received no external funding.

Institutional Review Board Statement: Not applicable.

Data Availability Statement: The data presented in this study are available on request from the corresponding author.

Acknowledgments: The authors would like to thank Li, Ge and Luo for their support.

Conflicts of Interest: The authors declare no conflict of interest.

References

1. Boshra, I.K.; Elbeih, A.; Mostafa, H.E. Composite Solid Rocket Propellant Based on GAP Polyurethane Matrix with Different Binder Contents. *Chin. J. Explos. Propellants* **2020**, *43*, 6.
2. Frankel, M.B.; Grant, L.R.; Flanagan, J.E. Historical development of glycidyl azide polymer. *J. Propuls. Power* **1992**, *8*, 560–563. [CrossRef]
3. Dey, A.; Sikder, A.K.; Talawar, M.B.; Chottopadhyay, S. Towards new directions in oxidizers/energetic fillers for composite propellants: An overview. *Cent. Eur. J. Energetic Mater.* **2015**, *12*, 377–399.
4. Selim, K.; Yilmaz, L. Thermal characterization of glycidyl azide polymer (GAP) and GAP-based binders for composite propellants. *J. Appl. Polym. Sci.* **2000**, *77*, 538–546. [CrossRef]
5. Li, M.-M.; Hu, R.; Xu, M.-H.; Wang, Q.-L.; Yang, W.-T. Burning characteristics of high density foamed GAP/CL-20 propellants. *Def. Technol.* **2021**, *18*, 1914–1921. [CrossRef]
6. Xu, M.; Ge, Z.; Lu, X.; Mo, H.; Ji, Y.; Hu, H. Fluorinated glycidyl azide polymers as potential energetic binders. *RSC Adv.* **2017**, *7*, 47271–47278. [CrossRef]
7. Shedge, M.; Tadkod, S.; Murthy, G.; Patel, C. Polyvinyl Acetate Resin as a Binder Effecting Mechanical and CombustionProperties of Combustible Cartridge Case Formulations. *Def. Sci. J.* **2008**, *58*, 390–397. [CrossRef]
8. Robbins, F.; Colburn, J.; Zoltani, C. *Combustible Cartridge Cases: Current Status and Future Prospects*; Army Ballistic Research Lab.: Aberdeen Proving Ground, MD, USA, 1992.
9. DeLuca, P.L.; Williams, J.C. Fibrillated polyacrylic fiber in combustible cartridge cases. *Ind. Eng. Chem. Prod. Res. Dev.* **1984**, *23*, 438–441. [CrossRef]
10. Stacer, R.G.; Husband, D.M. Molecular structure of the ideal solid propellant binder. *Propellants Explos. Pyrotech.* **1991**, *16*, 167–176. [CrossRef]
11. Xu, M.; Ge, Z.; Lu, X.; Mo, H.; Ji, Y.; Hu, H. Structure and mechanical properties of fluorine-containing glycidyl azide polymer-based energetic binders. *Polym. Int.* **2017**, *66*, 9.
12. Fataliev, R.V.; Eneikina, T.A.; Kozlova, L.A.; Gatina, R.F.; Mikhailov, Y.M. Study of an effect of promising cellulose components on the physico-mechanical properties and a burning rate of a combustible material for combustible cartridge cases produced according to the filtration casting technology. *J. Phys. Conf. Ser.* **2020**, *1666*, 012054. [CrossRef]
13. Zhao, Y.; Zhang, X.; Zhang, W.; Fan, X.; Xie, W.; Xu, H.; Liu, Y.; Du, J. GAP Factors affecting the mechanical properties of gap binder films. *Chin. J. Explos. Propellants* **2016**, *39*, 79–83.
14. Ma, S.; Li, Y.; Li, Y.; Li, G.; Luo, Y. Research on the Mechanical Properties and Curing Networks of Energetic GAP/TDI Binders. *Cent. Eur. J. Energetic Mater.* **2017**, *14*, 708–725. [CrossRef]
15. Chen, S.; Shi, H.; Tian, S.; Li, Z.; Xiao, Z. Design and fabrication of a composite coating to improve the environmental adaptability of combustible cartridges. *Polym. Compos.* **2018**, *40*, 685–694. [CrossRef]
16. Yang, Z.; Long, X.-P.; Zeng, Q.-X. Simulation study of the morphologies of energetic block copolymers based on glycidyl azide polymer. *J. Appl. Polym. Sci.* **2012**, *129*, 480–486. [CrossRef]
17. Yuan, S.; Jiang, S.; Luo, Y. Cross-linking network structures and mechanical properties of novel HTPE/PCL binder for solid propellant. *Polym. Bull.* **2020**, *78*, 313–334. [CrossRef]
18. Haska, S.B.; Bayramli, E.; Pekel, F. Adhesion of an HTPB-IPDI-based liner elastomer to composite matrix and metal case. *J. Appl. Polym. Sci.* **1997**, *64*, 2355–2362. [CrossRef]
19. Kakade, S.D.; Navale, S.B.; Narsimhan, V.L. Studies on Interface Properties of Propellant Liner for Case-Bonded Composite Propellants. *J. Energetic Mater.* **2003**, *21*, 73–85. [CrossRef]

MDPI AG
Grosspeteranlage 5
4052 Basel
Switzerland
Tel.: +41 61 683 77 34

Polymers Editorial Office
E-mail: polymers@mdpi.com
www.mdpi.com/journal/polymers